Annual Research Reviews

OXYTOCIN

Volume 2

M.Y. Dawood
Division of Reproductive Endocrinology
Department of Obstetrics & Gynecology
University of Illinois College of Medicine
Chicago, Illinois

OXYTOCIN
Volume Two
M.Y. Dawood

EDEN PRESS
Editorial Office
4626 St. Catherine St. West
Montréal, Quebec, Canada H3Z 1S3

AGENTS
University of Toronto Press
33 East Tupper Street, Buffalo, N.Y. 14203
University of Toronto Press
5201 Dufferin Street, Downsview, Ontario M3H 5T8
Eurospan
3 Henrietta Street, London WC2E 8LU, England
Australia & New Zealand Book Co.
10 Aquatic Drive, Frenchs Forest N.S.W. 2086 Australia

ISBN: 0-88831-112-5

Printed and bound in Canada by Les Editions Marquis

Dépôt légal – deuxième trimestre 1984
Bibliothèque nationale du Québec

TABLE OF CONTENTS

Chapter Seven – OXYTOCIN RECEPTORS

Chapter Eight – OXYTOCIN-NEUROPHYSIN INTERACTION AND OXYTOCIN METABOLISM

Chapter Nine – THE OXYTOCIN CHALLENGE TEST

Chapter Ten – INDUCTION OF ABORTION AND LABOR

Chapter Eleven – OXYTOCIN AND THE FETUS/NEONATE

Chapter Twelve – OXYTOCIN ANALOGS

≈≈ To my daughter, Fatimah,
a beneficiary of oxytocin

PREFACE

This research review covers the research publications on oxytocin which have appeared in the *Index Medicus* for the four years from January 1977 to December 1980. More than 95% of the publications in English were reviewed. A few were omitted because of inaccessibility or because they were mainly review articles that were based on other original publications. Most of the publications not in English have been included except for a few readily available ones which had an abstract in English. The papers have been arranged into twelve chapters. Material from some publications spanned more than one chapter. Others could be placed exclusively into one or the other of twelve chapters. Each chapter starts with a brief review, in a few lines, of what was generally accepted on that topic together with the areas of research publications that have appeared during the four-year period. Where indicated, the author has inserted his critical analysis of the particular publication.

This book would not have been produced without the assistance of many. The excellent resources and facilities of the library at the University of Illinois Medical Center in Chicago eased the search and accessibility to the relevant publications. The skillful and diligent help of my secretary, Ms. Mary Jackson, speeded up the evolution of the manuscript. Special thanks to my wife, Dr. Firyal S. Khan-Dawood, and my daughter, Fatimah, who accepted many evenings and weekends of neglect.

The conducive academic milieu of the Department of Obstetrics and Gynecology, University of Illinois College of Medicine under the chairmanship of Dr. W.N. Spellacy, encouraged the writing of this book. Last, but not least, the patient tolerance of Dr. David Horrobin, series editor, is much appreciated. To those whose publications on oxytocin appeared in the *Index Medicus* 1977-1980 but were inadvertently not included in this series, I offer my apologies.

M.Y. Dawood, M.D.
Division of Reproductive Endocrinology
Department of Obstetrics & Gynecology
University of Illinois College of Medicine
Chicago, Illinois

Chapter 1

▪ SYNTHESIS AND LOCAL TRANSPORT

DISTRIBUTION OF OXYTOCIN NEURONS

Immunocytochemical localization

van Leeuwen (363) has reviewed the immunocytochemical specificity for oxytocin and vasopressin. Apparently nonspecific staining is a major problem with immunocytochemistry, although specificity has been cited as the main feature of this technique. Absorption of the antiserum by the respective hormone could be employed to demonstrate method specificity, but not serum specificity. Characterization of the different antigens in the tissue may therefore be necessary. Many of the conventional methods employed for characterization of proteins are not suitable for neurohypophysial hormones because of their small size. Isoelectric focusing cannot distinguish arginine vasopressin from arginine vasotocin since they have the same isoelectric point. The removal of cross-reacting antibodies is not always possible. For example, van Leeuwen (363) found that antibodies he had raised in his institution against vasotocin cross-reacts with vasopressin, but that solid-phase absorption to vasopressin resulted in the removal of all the antibodies against vasotocin, too. The purified antiserum could be tested against tissue known not to have the antigen and thus the antisera should give negative staining. For this purpose the Brattleboro rat (which is lacking in vasopressin) is an ideal source of tissues with no vasopressin. Radioimmunoassay could be employed for evaluating specificity and titer of the antiserum but it is still not indicative of the potency and specificity of the antiserum for immunocytochemistry since there may be marked differences in the relative concentration of antigens and antibodies used in the two techniques. Finally, by virtue of their high degree of specificity and high titers, monoclonal antibodies might overcome some of the difficulties of immunocytochemistry. However, false negative results may be a problem because the antibodies often possess one type of combining site and the antigenic determinant may be masked *in vivo*. Perhaps this could be overcome by using antibodies from different clones to study the same tissue.

Hypothalamo-Neurohypophysis

Immunocytochemistry has been employed to localize the oxytocin-containing neurons and the distribution of oxytocin in different regions of the brain and even the spinal cord. The technique has been used to localize the distribution of oxytocinergic neurons in different animal species such as the rat, cat, frog and reptiles, and showed that oxytocin, vasopressin and their neurophysins were found in the magnocellular cell bodies of the hypothalamus with their axonal projections to the neural lobe of the pituitary (410). The Brattleboro rat, which is genetically deficient in arginine vasopressin and has diabetes insipidus, has been used very conveniently to probe the distribution of neurohypophyseal hormone neurons. The evidence from this unique rat confirmed that the one neuron, one hormone concept for oxytocin and vasopressin was applicable to the rat (234).

The observation that vasopressin and oxytocin were synthesized in separate neurosecretory neurons in mammals has been extended to include the human hypothalamus. Employing specific antiserum which had been absorbed to cross-reacting neurohypophyseal hormones conjugated to Sepharose-B, immunocytochemical localization of vasopressinergic and oxytocinergic neurons in the adult human hypothalamus confirmed that oxytocin and vasopressin were synthesized in separate neurons (103). Two separate types of neurons were selectively stained but no neuron showed "double" staining for both hormones. The supraoptic nucleus was mainly dominated by vasopressinergic neurons, 95% being vasopressinergic in the dorsolateral part and 80-90% in the medial part of the nucleus. In the intersupraoptico-paraventricular islands, the number of oxytocinergic neurons was relatively high. In the paraventricular nucleus, the mass was predominated by vasopressinergic neurons (70-80%) but in the filiform cell aggregation anteriorly and in the roof of the optic recess, oxytocinergic neurons predominated (80%). Thus, oxytocinergic neurons had a preferential location at the periphery of the paraventricular nucleus. In another study, human fetal hypothalamus was shown by immunocytochemistry to possess oxytocin-neurophysin neurons, the perikarya of which were located in paraventricular and supraoptic nuclei (261). The median eminence also had immunoreactive neurons. Neurophysin neurons were more numerous than oxytocin neurons and the first neurophysin-oxytocin neurons could be demonstrated as early as the fourteenth week of gestation.

Both the supraoptic and paraventricular nuclei of the human hypothalamus contained at least two different types of neurophysins, one associated with vasopressin and the other with oxytocin (104). The human oxytocin and vasopressin associated neurophysins were immunologically related to bovine neurophysin I and neurophysin II respectively, since they reacted with the respective bovine anti-neurophysin antisera immunocytochemically. Human vasopressin-neurophysin and oxytocin-neurophysin were located separately in two different types of neurons, which corresponded respectively to the vasopressinergic and oxytocinergic neurons of both the supraoptic and paraventricular nuclei. The neurophysin found in the human vasopressinergic suprachiasmatic neurons appeared to be similar or identical to the neurophysin of the vasopressinergic neurons of the human magnocellular hypothalamic nuclei.

Zimmerman et al. (408, 409) described the magnocellular neurosecretory pathways in the monkey, human adult and rats, using immunohistochemistry. In the rhesus monkey, neurohypophyseal hormones and their respective neurophysins were localized in the magnocellular neurons of the supraoptic nucleus and the paraventricular nucleus. Estrogen-stimulated neurophysin was present in oxytocinergic neurons. Unlike vasopressin, perikarya containing oxytocin and estrogen-stimulated neurophysin tended to be concentrated in the dorsal region of the supraoptic nucleus and scattered in different regions of the paraventricular nucleus.

In the cat, the organization of the magnocellular neurosecretory neurons in the hypothalamus was essentially similar to that described in other mammals except that the cat hypothalamus is uniquely endowed with the lateral hypothalamic accessory nucleus (280). Vasopressin- and oxytocin-containing neurons were found in the supraoptic, the paraventricular and in five accessory nuclei. These five accessory nuclei are the circularis nucleus, the anterior fornical nucleus, the posterior fornical nucleus, the retrochiasmatic nucleus and the lateral hypothalamic nucleus. Both vasopressinergic and oxytocinergic cells were equally mixed in the paraventricular and the five accessory nuclei but the supraoptic nucleus was dominated mainly by vasopressinergic neurons.

In the pig too, the one hormone-one neuron and one hormone-one neurophysin concept holds true (384). Oxytocin and lysine vasopressin were specifically localized by immunocytochemistry in the paraventricular and supraoptic nuclei of the pig hypothalamus. Oxytocinergic neurons dominated the caudal part of the supraoptic nucleus while both oxytocinergic and vasopressinergic neurons were uniformly present in the rostral part. In the paraventricular nucleus, both peptides were evenly spread in the rostral portion and also close to the third ventricle. In the caudal part of the paraventricular nucleus, oxytocin neurons were mainly adjacent to the third ventricle and vasopressin neurons lateral to this and in the dorsocaudal area. In the paraventricular nucleus, distribution of oxytocin and vasopressin paralleled that of porcine neurophysin II and neurophysin I respectively.

Arginine vasopressin (AVP) fibers tend to cluster in the central region of the neural lobe, entering it bilaterally, while oxytocin fibers were found mainly in the peripheral part (364). These findings questioned previous reports of AVP in the pars intermedia cells since similar staining of the pars intermedia with anti-AVP was obtained with Wistar and homozygous Brattleboro rats, mutant strains deficient in AVP.

With the Wistar and Brattleboro rats, marked cross-reactivity between antivasopressin and antioxytocin preparation was found if proper precautions were not taken (365). If the cross-reactivity of the antiserum was eliminated by sufficient dilution or by absorption to agarose beads coated with the cross-reacting component, then the neurohypophysis of the Wistar rats revealed two types of neurosecretory fibers distributed in clusters. Staining of the pars intermedia cells of the rat with anti-AVP could be removed by solid-phase absorption to α-melanocyte stimulating hormone (α-MSH) and was most probably due to cross-reaction with α-MSH. Somatostatin fibers appeared to be present in the peripheral parts of the proximal neurohypophyseal stalk and mainly lateral in its more distal parts, with a rapid decrease in number in the neural lobe. Magnocellular elements were probably the result of the cross-reaction of antisomatostatin with neurophysins. Thus it appeared that the rat neural lobe had many oxytocinergic and vasopressinergic fibers together with neurophysins and that the intermedia probably had little or no AVP.

Extrahypothalamo-neurohypophyseal Distribution

The cells of the magnocellular neurosecretory system were shown to be more widely distributed than previously suspected (184). Using horse radish peroxidase injection into the rat posterior pituitary lobe, oxytocin-containing or both oxytocin- and vasopressin-containing neurons were found in the caudal part of the red nucleus of the stria terminalis, in the median and the periventricular preoptic nuclei, in the medial and lateral preoptic areas, in the zona incerta and in the substantia innominata.

Using purified first antiserum in the immunohistochemical method, vasopressin- and oxytocin-containing pathways in the rat could be traced from the paraventricular nucleus towards the dorsal and ventral hippocampus to the lateral ventricle, the stria terminalis, the stria medullaris, the nuclei of the amygdala, substantia nigra and substantia grisea, nucleus tractus solitarius, nucleus ambiguus and to the substantia gelatinosa of the spinal cord (53, 54, 57). Only vasopressin-containing fibers ran from the parvocellular suprachiasmatic nucleus to the organum vasculosum laminae terminalis, lateral septum and the lateral habenular nucleus (54). However, the physiological function of the oxytocin and vasopressin fibers is not known. These fibers may be involved in the maintenance of water balance, milk ejection, and in the learning and memory processes such as the retention of passive avoidance behaviour. Another suggestion is that both oxytocin and vasopressin may be putative neurotransmitters.

Buijs and Swaab (56) pursued their previous light microscopy observations that oxytocin and vasopressin nerve fibers terminated on other neurons to demonstrate the existence of such fibers in the limbic system, which is one of the main target areas for these peptides. The brains of male Wistar and Brattleboro rats, homozygous for diabetes insipidus, were studied immunoelectron microscopically with and without osmium tetroxide (OsO_4). Post-embedding staining produced false positive reaction on all dense core vesicles. With pre-embedding staining, intense and specific reactions were seen for both vasopressin and oxytocin at their sites of production, the neurohypophysis and in the extrahypothalamic limbic regions. Within the limbic system, only vasopressin-containing synapses were seen in the lateral septum and habenular nucleus but the medial nucleus of the amygdala synapses contained either vasopressin or oxytocin. These neurohypophyseal peptide-containing synapses did not appear to differ in any fundamental way from the classical transmitter-containing synapses in the brain.

Electron microscopic immunocytochemical technique was also used to confirm light microscopy findings that in the rat, the neurohypophyseal hormones were located in separate neurophysin-vasopressinergic and neurophysin-oxytocinergic nerve fibers (13).

The relationship of oxytocin and luteinizing hormone releasing hormone (LRH) to the circumventricular organs of the rat and guinea pig brain was determined (388). Oxytocin fibers were found to project to the neural lobe but vasopressin fibers were found in the neural lobe as well as the zona externa of the median eminence. In the organum vasculosum laminae terminalis (OVLT), LRH, oxytocin and somatostatin terminals were found around the blood vessels, indicating a neurohemal response. The OVLT of the guinea pig was found to be immediately surrounded by magnocellular, mainly oxytocin-containing perikarya, suggesting again a possible hemoneural relationship.

The subfornical organ and area postrema had very few neurohormone-containing fibers while the commissural organ and the pineal organs had none.

Extrahypothalamic neurophysin-oxytocin fibers were found projecting to the brain stem and spinal cord but did not interact with the area postrema.

Oxytocin neurons have projections to the limbic system, diencephalon, mesencephalon, brain stem and spinal cord in the rat (332). In the human, the brain stem and spinal cord showed oxytocin and vasopressin fibers projecting from the hypothalamic magnocellular neurons and vasopressin fibers from the parvocellular neurons of the suprachiasmatic nucleus. Both axosomatic and axodentritic contacts were present in these neural target areas. Such observations suggested that these projections interacted with other neurons rather than released hormone into the blood stream.

Effect of Stimulus on Oxytocin Fibers

In vertebrates, the two neurohypophyseal hormones are synthesized in two different types of endocrine neurons. Vandesande and Dierickx (367) wanted to determine if this was also the case when the animal was subjected to intense and prolonged stress. Rats were subjected to dehydration and dehydration combined with estrogen, both of which were known to stimulate neurohypophyseal hormone release. Using immunocytochemistry with the unlabeled antibody peroxidase-antiperoxidase complex (PAP) technique, the results indicated that in rats with activated hypothalamic magnocellular neuroendocrine systems, vasopressin and oxytocin were exclusively synthesized in separate vasopressinergic and oxytocinergic neurons.

In another study using specific immunocytochemistry, normal and Brattleboro rats with diabetes insipidus were studied for distribution of the oxytocin and vasopressin neurons under basal conditions and with osmotic stimulation during dehydration (68). The findings confirmed that oxytocin and its neurophysin, and vasopressin and its neurophysin were synthesized in different neurons and transported along different axons. The supraoptic and paraventricular nuclei were functionally indistinguishable since both hormones were distributed in both nuclei. Both types of neurons responded to osmotic stimulation in a similar qualitative way but differed quantitatively. More detailed analysis of the hypothalamic nuclei revealed that during prolonged thirst in the rats, the posterior supraoptic nucleus and the accessory cell clusters reacted more markedly than the anterior supraoptic nucleus (193). In the median eminence, the neurosecretory axons showed close contacts with the portal vessels not only in its lateral portion but, in dehydrated animals, also around the midline.

The hypothalamo-infundibular tract of the rat carrying vasopressin and neurophysin II is involved in the regulatory mechanism of the pituitary-adrenal axis but oxytocin does not appear to play a significiant role (59). Oxytocin and neurophysin I fibers were rarely distributed in the external layer of the median eminence and this was unchanged after adrenalectomy or dexamethasone therapy.

In contrast to these immunocytochemical observations, measurements of oxytocin concentration in specific hypothalamic nuclei indicated that oxytocin may play a role in sexual maturation. Weanling rats were found to have a significantly higher concentration of oxytocin in the supraoptic nucleus, median eminence and retrochiasmatic area than adult rats (178). The vasopressin concentrations were similar. Significantly greater concentrations of oxytocin were also found in adult female rats in the paraventricular nucleus, suprachiasmatic nucleus, anterior hypothalamic nucleus and posterior pituitary compared to the weanling rats. These differences were thought to suggest a role for sexual maturation in the increasing synthesis and storage of oxytocin in specific hypothalamic nuclei and in the posterior pituitary respectively.

Vasotocin and Mesotocin Neurons

The hypothalamic magnocellular neurosecretory system of lizards studied with the unlabeled antibody peroxidase-antiperoxidase complex (PAP) technique at the light microscopic level revealed that vasotocin and mesotocin were synthesized in separate neurons and their perikarya were of different sizes (153). Although both cell types were in close juxtaposition, they did not have a distinct pattern of distribution except for the external zone of the median eminence which contained numerous vasotocinergic fibers but few mesotocinergic fibers.

Previous speculation that the teleosts synthesized their two neurohypophyseal hormones, vasotocin and isotocin, in separate neurons within the preoptic nucleus was recently confirmed (154). Vasotocin and isotocin were immunocytochemically demonstrated to be in different perikarya and the neurohypophysis contained separate vasotocinergic and isotocinergic nerve fibers.

Electron microscopy has been employed to assist with studies attempting to map the distribution of oxytocinergic fibers by immunocytochemistry. This technique was used and confirmed an earlier observation with light microscopy that the internal and external regions of the median eminence of the frog contained separate vasopressinergic and mesotocinergic nerve fibers (366). It was found that the median eminence contained separate vasopressinergic and mesotocinergic nerve fibers, and the means size of the neurohypophyseal hormone-containing granules in the external region of the median eminence was significantly smaller than those in the internal region. However, there was no difference in the granule size of the vasotocinergic and mesotocinergic fibers throughout the median eminence.

Oxytocin Fibers in the Spinal Cord

With the aid of immunohistochemical methods, oxytocin-stained and neurophysin-stained fibers have been demonstrated to be present in the spinal cord of the rat and the monkey (347). In albino rats, oxytocin-stained fibers descended through the dorsal part of the lateral funiculus to the caudal end of the cord. Fibers left the lateral funiculus to innervate the marginal zone of the dorsal horn at all levels, and the intermediolateral column at thoracic, lumbar and sacral levels. The central grey matter was innervated with oxytocin-stained fibers. In the monkey, oxytocin-stained fibers descended also through the lateral funiculus to the filum terminale. The intermediolateral column was not innervated at T_1-T_3, moderately innervated at T_4-L_3, and sparsely innervated at the sacral levels. Very few or no oxytocin-stained fibers were present in the central gray and marginal zone. The Brattleboro rat also had oxytocin-stained and neurophysin-stained fibers in the intermediolateral column, central gray and marginal zone. It appeared that the paraventriculo-spinal pathway was particularly related to specific groups of sympathetic and parasympathetic preganglionic neurons, and to the marginal zone, which was involved in transmission of ascending nociceptive information through the spinothalamic tract.

The Pineal Gland

Pavel and his group had proposed, and it was widely accepted, that arginine vasotocin was the peptide hormone of the mammalian pineal gland. However, there continued to be controversy about the existence of vasotocin in the pineal and pituitary glands of mammalian and nonmammalian species. Since then there have been a number of studies suggesting that arginine vasopressin rather than vasotocin could be present in pineal gland and extrahypothalamic neurons and fibers. Therefore, some investigators have employed radioimmunoassay (RIA) determination of these neurohypophyseal peptides in brain tissues using specific antiserum. Using sensitive radioimmunoassays, oxytocin, arginine vasopressin and vasotocin contents of the chicken pituitary and pineal glands, bovine and rodent pineal glands were compared (128). In the chicken pituitary and pineal glands, only vasotocin (up to 1.48 mg/pituitary gland and 300 pg/pineal gland) was detectable. Rodent pineal glands also contained only vasotocin but in lower concentrations (10 pg/gland). In contrast, bovine pineal gland contained all three peptides in approximately similar concentrations (about 200-400 pg/gland). Since vasotocin was immunologically detectable in the pineal of several species, it was suggested that it might subserve some physiologic function such as an antigonadotropic hormone in mammals. The antiserum for vasotocin used in this study did show cross-reactivity with both oxytocin and vasopressin. With more specific antiserum for vasotocin, several studies have concluded that the previously reported presence of vasotocin was due to poor specificity of the antiserum used.

Thus, Dogterom et al. (110) who used a specific radioimmunoassay for vasotocin, were unable to demonstrate the presence of immunoreactive vasotocin in the pineal gland and subcommissural organ of mammalian species. Measurements of oxytocin, arginine vasopressin and vasotocin by specific radioimmunoassays showed that in the rat and bovine pineal glands, vasotocin was undetectable but oxytocin and vasopressin were present in small amounts, except in the pineal of the Brattleboro rat where vasopressin was undetectable (109). Unlike the rat subcommissural organ, rabbit subcommissural organs had detectable amounts of vasotocin. The investigators concluded that a vasotocin-like compound could be present in mammalian pineals and/or subcommissural organs but that this compound was different from vasotocin itself.

Direct measurements of pituitary gland homogenates revealed the presence of vasotocin only in the frog and chicken pituitary glands (nonmammalian), and not in midterm and near-term fetal seal (mammalian) although oxytocin and vasopressin were present in the fetal seal pituitary (110). While oxytocin and vasopressin could be measured by RIA in rat and bovine pineal glands and in the subcommissural organs of rats and rabbits, specific vasotocin immunoreactivity was not detectable. This discrepancy between the presence of biologic activity of vasotocin, but absence of immunologic activity of the peptide in the pineal and pituitary glands and the subcommissural organ was attributed to the presence of a peptide which was structurally closely related to, but not identical with vasotocin.

Vasopressin and oxytocin fibers extended via the subcommissural organ or habenular commissure into the pineal stalk and terminated in the anterior part of the pineal organ (55). Vasopressin-containing fibers also run caudally adjacent to the subcommissural organ towards the central grey matter. The authors were therefore able to show that vasopressin and oxytocin were present in the stalk and anterior part of the pineal gland and were connected by oxytocinergic and vasopressinergic fibers via the subcommissural organ and habenular commissure. It is possible that what was ascribed to vasotocin in the pineal gland was probably due to vasopressin, since definitive vasopressinergic fibers connecting the hypothalamo-neurohypophyseal system to the extra-hypothalamic fibers and pineal gland were demonstrable.

BRAIN NEUROHYPOPHYSEAL HORMONE CONTENT

Microdissection of the hypothalamic nuclei and surrounding neural tissues have been carried out and the oxytocin and vasopressin content of the various nuclei determined by RIA. This has contributed to our understanding of the distribution of both hormones in the hypothalamo-neurohypophyseal system.

In the homozygous Brattleboro rats, microdissected hypothalamic and the posterior pituitary had virtually no vasopressin as determined by RIA, but had normal concentrations of oxytocin except for the arcuate nucleus (148).

Adult male Zucker rats, which are genetically obese, had significantly lower concentrations of oxytocin in their pituitary glands and median eminence compared to phenotypically normal lean rats (85). The changes were similar to those seen for vasopressin concentrations. The levels of oxytocin in the retrochiasmatic area and the paraventricular and anterior hypothalamic nuclei were similar in obese and nonobese rats. Increased noradrenergic transmission in the paraventricular nucleus may be responsible for the changes seen with vasopressin but no explanation was offered for oxytocin. It is unclear whether the changes in pituitary and median eminence oxytocin in the obese rat were the result of oxytocin release secondary to osmotic changes in the plasma (Chapter 2) or of a central adrenergic mechanism (Chapter 2) or a combination of both.

Measurements of oxytocin and vasopressin by RIA in individual hypothalamic nuclei obtained from ten human subjects who died suddenly while in good health, revealed that both oxytocin and vasopressin were present in appreciable amounts in both the paraventricular and supraoptic nuclei (147). Similar concentrations of vasopressin were found in the supraoptic (440.0 ± 156 pg/mg protein) and paraventricular nuclei (393.0 ± 158 pg) with lower concentrations in the arcuate, dorsomedial, ventromedial and posterior hypothalamic nuclei, median eminence and mamillary bodies. The pituitary stalk had even higher concentrations of vasopressin (1577.0 ± 66 pg) than the supraoptic and paraventricular nuclei. In contrast, the supraoptic nucleus had less oxytocin (28.0 ± 4.9 pg/mg protein) than the paraventricular nucleus (52.0 ± 7.9 pg/mg protein). It should be noted that while the results are qualitatively useful, postmortem changes could influence the quantitative relationship.

SYNTHESIS AND AXONAL TRANSPORT

The synthesis, transport and release of posterior pituitary hormones were investigated using pulse labeling of neurons of the hypothalamo-neurohypophysial system with ^{35}S-cysteine (49, 140). The findings indicated that oxytocin, vasopressin and their respective neurophysin carrier proteins were synthesized as parts of separate precursor proteins. Using this pulse labeling technique in normal and Brattleboro rats, it was found that both the vasopressin-associated neurophysin (Np-AVP) and the oxytocin-association neurophysin (Np-OT) were synthesized as larger precursor protein molecules in the perikarya of the neurons of the supraoptic nucleus. A precursor-to-product conversion then took place within the neurosecretory granules during axonal transport of the precursors. Isoelectric focusing showed the precursor forms for Np-AVP and Np-OT had isoelectric points (pI) of 6.1 and 5.4 respectively and gave rise to the intermediate forms with a pI of 5.6 and 5.1 respectively. Finally, the intermediate form underwent conversion to the final product with a pI of 4.8 and 4.6 respectively for Np-AVP and Np-OT (molecular weight 10,000-12,000 daltons) which were the major stored products in the posterior pituitary gland. When chromatographically purified Np-AVP precursor was trypsinized, a 10,000 dalton material (pI = 4.8) which bound to vasopressin and peptides were formed. The peptides bound to neurophysin-sepharose, the binding being enhanced if the particles had prolonged incubations with trypsin. With the precursor Np-OT, trypsinization produced neurophysin-like and oxytocin-like molecules. It was proposed that the vasopressin Np-AVP precursor be called *pro-pressophysin* and the oxytocin Np-OT precursor be called *pro-oxyphysin.* It was assumed that all the necessary enzymes, including carboxamidopeptidase, were present in the neurons to convert these precursors to neurophysin and the respective neurohypophyseal hormones. Pro-pressophysin was shown to be a glycoprotein, whereas pro-oxyphysin did not appear to have a carbohydrate in its molecule. In the Brattleboro rats, there was a third neurophysin precursor which was not a glycoprotein but comigrated with pro-pressophysin on Sephadex G-75. Tryptic digestion gave a neurophysin-like molecule and peptide products which were different from those of pro-pressophysin and pro-oxyphysin trypsinization. This new neurophysin precursor could be a vasotocin precursor or a nonglycosylated vasopressin precursor.

Previous studies have shown that colchicine, a mitotic poison which bound to microtubule protein and inhibited transport in the neuronal system, was able to inhibit intraneuronal transport of oxytocin-neurophysin and vasopressin-neurophysin when injected in doses of 40-80 μg into rat brain. Parish and Pickering (258) demonstrated that similar effects were accomplished with lower doses of colchicine, up to 20 μg/animal. Using ^{35}S-cysteine injected intracisternally and which was incorporated into the neurophysins, colchicine in doses of more than 5 μg/animal decreased intraneuronal transport of oxytocin-neurophysin and vasopressin-neurophysin to 10% or less, but a dose of 2 μg/animal had no significant effect. With 3.5 μg colchicine, the reduction of intraneuronal transport for neurohypophysial peptides was reduced to 40%. The duration of block with the 3.5 μg dose was longer for vasopressin-neurophysin containing neurons than for oxytocin-neurophysin ones with the peak effect seen three days after colchicine and returning to normal by six days. The results suggested a differential effect of colchicine on these two neuronal systems with the effect of colchicine being greater on the vasopressin-containing neurons than the oxytocinergic neurons.

Monosodium glutamate, a chemical used in cooking and many food products, reduced protein content in the arcuate nucleus and median eminence in ten-day-old rats and in the arcuate and suprachiasmatic nuclei of adults (303). Oxytocin and vasopressin content of the posterior pituitary of the young rat remained unchanged. However, in adults, monosodium glutamate caused a depletion of oxytocin and vasopressin in the paraventricular nucleus, and depletion of vasopressin in the supraoptic nucleus without any change in the oxytocin content. Changes in the hormone content of the hypothalamic nuclei were more variable. It was suggested that increased hormone content may reflect accumulation due to reduced transport along the axons while decreased content may indicate suppression of synthesis. Thus monosodium glutamate could potentially suppress synthesis and axonal transport of neurohypophyseal hormones.

■ ■ ■

Chapter 2

▪ SECRETION

Neurohypophyseal hormones are known to be stored intracellularly in the form of granules which are released from the cell by a process of exocytosis. There is some debate as to whether the membrane of the granule is recaptured by the cell. Electrophysiologic alterations in the hypothalamic nuclei have been shown to occur during secretion of oxytocin. Several factors are known to stimulate release of oxytocin, the better known ones being nipple-suckling during lactation and the Ferguson reflex. Several less established factors for stimulating oxytocin secretion have been examined recently. This chapter will review the recent literature on investigations into the mechanism controlling secretion of oxytocin.

CELLULAR MECHANISM

The mechanism by which neurohypophyseal hormones are released from the perikarya has been studied; the method employed was examination of the morphological changes of the cell membrane. Polypeptide-secreting cells, including oxytocin- and vasopressin-secreting neurons, are known to release materials stored in membrane-bound granules through a process called exocytosis. Exocytotic release of the secretory products occurs when the granule-limiting membrane fuses with the cell membrane. Such a fusion of the membrane occurs in protein-depleted segments of the respective membranes since a displacement of the intramembranous particles from the prospective fusion sites is noted. Theodosis et al. (354) examined the morphology of membrane changes during neurohypophyseal hormone release in hibernating garden dormice. Hibernation was artificially induced in these rodents by placing them at 4°C, and reawakening was induced by tactile stimulation. The authors were able to show that the release of hormones was reflected morphologically in the frequent incidence of exocytotic profiles in the neurohypophyseal axon terminals of awakening animals examined with conventional and freeze-fracture electron miscroscopy. There was a marked increase in plasma vasopressin from

37 ± 18 mU/ml (mean ± SEM) at hibernation to 279 ± 80 μU/ml at 30 minutes and 712 ± 224 μU/ml at 3 hours. At the same time, thin sections of neurohypophyseal axons revealed plasmalemma invaginations containing electron-dense secretory granule core substance exposed to extracellular space, representing the process of exocytosis. In fully awake dormice, such plasmalemma invaginations were rare, but two or more of them were often obvious in one axon terminal from awakening animals. Freeze-fracture preparations revealed very few or no particles over an extensive area of the membrane surrounding the exocytotic openings and from the leaflet limiting the surface invaginations which represented the granule membrane fused to the plasmalemma. These observations supported the earlier observations of membrane-associated changes during neurohypophyseal hormone release in other animal species and confirmed a similar process in the rodent.

ELECTROPHYSIOLOGIC DETERMINANTS

Neurosecretory cells of the paraventricular and supraoptic nuclei can be identified by antidromic stimulation and by recording electrophysiological activity during reflex milk ejection and hemorrhage in the lactating rats (117, 275). The simplest method was to apply stimuli which specifically released only neurohypophyseal hormone while making a single unit recording from the nuclei. Using this method, the paraventricular and supraoptic neurosecretory cells can be electrophysiologically differentiated into two distinct populations; 1) oxytocin-releasing neurons which showed a high-frequency discharge before milk ejection induced by suckling and 2) vasopressin-releasing neurons which adopted a phasic pattern of firing during vasopressin release induced by hemorrhage. It was suggested that the rate of vasopressin secretion into the circulation depended largely on the proportion of vasopressin neurons firing phasically, their firing rates within the phases, and the duration and degrees of synchronization of the phases. The mechanism of phasic firing may be through an oscillating input from sources outside the paraventricular and supraoptic nuclei, through a local feedback control circuit within the magnocellular nuclei or may be an inherent property of the membranes of the neurosecretory cells, the latter being the mechanism favored by Dyball et al. (117). The intervals between the bursts represented a "recovery period" in which the ionic and structural integrity of the neurosecretory cells may be restored during a sustained stimulus.

In contrast to previous observations that less than 10 Hz of electrical stimulation of the pituitary stalk *in vivo* was ineffective in producing neurohypophyseal hormone release, a recent study showed that low-frequency stimulation of the pituitary stalk was indeed effective in releasing oxytocin *in vivo* (37). Electrical stimulation of the pituitary stalk with 6 Hz resulted in an increased plasma oxytocin concentration 30 minutes later, uterine contractions in the early postpartum rats, and milk ejection in the lactating rat.

REGULATORY MECHANISMS

Factors regulating the secretion of oxytocin from the posterior pituitary gland are only partially known. Some of the factors that are more widely known or accepted as capable of stimulating oxytocin release include nipple-suckling, stretching of the cervix and upper vagina (the Ferguson reflex), the expulsive phase of labor, and perhaps coitus. Other factors that have been associated with oxytocin release include administration of prostaglandins for induction of labor at term, estrogens, excessive hemorrhage in some species, thyrotropin, dopa, and electrical stimulation of some of the hypothalamic nuclei. Some of these factors influencing secretion were summarized at a recent meeting (166). Investigations have focused on stimuli or factors that could stimulate or attenuate oxytocin secretion and the mechanism involved in regulating oxytocin release.

Dopaminergic Control

The dopaminergic mechanism regulating oxytocin release in different psysiological states was examined by Seybold et al. (313) who demonstrated the ability of dopamine, levodopa and carbidopa to inhibit spontaneous release of oxytocin in male and pregnant rats. *In vitro* experiments demonstrated that dopamine inhibited the spontaneous release of oxytocin from the hypothalamo neurohypophyseal system of male rats in a dose-dependent manner. The maximum inhibition was observed with concentrations of 10^{-9} dopamine and for apomorphine. Haloperidol in concentrations of 5×10^{-5}M was able to block the effect of dopamine on oxytocin release. *In vivo* experiments showed that levodopa and carbidopa injected into male rats caused a significant increase in pituitary gland content of oxytocin. In pregnant rats at 16 days gestation, levodopa and carbidopa were able to block oxytocin release and delayed average delivery time by 12 hours. While the type of receptor involved in the dopa inhibition of oxytocin release was not established, the results of these investigators suggested a dopaminergic mechanism for the regulation of oxytocin release.

Clarke et al. (72) were able to demonstrate that during suckling in the rat, the neural pathway for the reflex release of oxytocin contained both dopaminergic and noradrenergic synapses, the latter acting through α-adrenoceptors and being distal in the pathway to the dopaminergic component. During suckling, anesthetized lactating rats have been shown to regularly release, every seven minutes, brief pulses of oxytocin (0.5 - 1.0 mU) which produce single transient increases in intramammary pressure. This model was then employed to evaluate the effects of various agents which selectively impaired synaptic transmission in order to determine the role of dopamine and noradrenaline in regulating this natural reflex.

The release of oxytocin was determined by measurement of the intramammary pressure. The suckling-induced release of oxytocin was blocked by:

1) inhibitors of catecholamine synthesis such as diethyldithicarbamate (100 - 200 mg/kg i.v.) and α-methyl parathyrosine (100 - 400 mg i.v.);
2) dopamine antagonists, fluphenazine (0.7 mg/kg), pimozide (1.4 mg/kg), cis-flupenthixol (4.5 mg/kg) and metoclopramide (6.0 mg/kg) in a dose-dependent manner;
3) the α-adrenoreceptor antagonist phenoxybenzamine (1.4 mg/kg).

In contrast, dopamine (40 mg), bromocriptine (10 mg), apomorphine (100 mg), noradrenaline (10 mg) and phenylephrine (2 mg) injected into the cerebral ventricles caused a sustained release of oxytocin with multiple rises in intramammary pressure. Isoprenaline (4 mg) was however, ineffective. Dopamine- and noradrenaline-induced release of oxytocin during lactation was prevented by cis-flupenthixol and phenoxybenzamine respectively. The mammary sensitivity to exogenous oxytocin remained unaffected by any of the drugs tested and the effect of these drugs was not affected by pretreatment with propranolol (1 mg/kg). These results therefore suggested that both dopaminergic and noradrenergic synapses were present in the neural pathway for the suckling induced reflex release of oxytocin.

Adrenergic Control

Adrenergic mechanisms may be involved in oxytocin release from the hypothalamo-neurohypophyseal system and may potentiate the effect of osmoreceptor stimulation in dehydrated animals (159). Dehydration in male rats caused a release of oxytocin. Injection of phenoxybenzamine hydrochloride, an alpha adrenergic receptor blocker and amphetamine, an alpha and beta adrenergic receptor stimulant, into nondehydrated and 12 days dehydrated male rats stimulated an increase in oxytocin release from the hypothalamo-neurohypophyseal system in nondehydrated animals and further, significantly potentiated the release of oxytocin in dehydrated rats. The finding of reduced hypothalamic content of oxytocin following dehydration was at variance with earlier observations that hypothalamic oxytocin content was increased in dehydrated rats but agreed with previous findings of reduced neurohypophyseal oxytocin content during dehydration. In spite of different adrenergic mechanisms, both phenoxybenzamine and amphetamine depleted oxytocin stores of the hypothalamo-neurohypophysis in dehydrated and nondehydrated rats, suggesting that suppression of alpha adrenergic receptors and stimulation of alpha and beta adrenergic receptors produced release of oxytocin.

That adrenergic mechanism were involved in the reflex release of oxytocin was confirmed also in lactating rats (232, 357). However, there was some variance in the facilitatory and inhibitory effects of the alpha and beta adrenergic receptors. Unlike the dehydrated rat, where the osmostimulatory effect had to be contended with also, in lactating rats the milk ejection itself can be used as an index of the effect of various alpha and beta adrenergic receptor blockers. Propranolol (1.0 - 1.5 mg/kg i.v.) and other beta antagonists had no effect on the milk ejection reflex in animals that were already milk-ejecting, but propranolol and oxyprenolol were able to promote milk ejection in rats that had not established or had ceased milk ejection (357). The L-isomer, but not the D-isomer, of propranolol was effective when given either intravenously or intraventricularly. Practolol, a beta antagonist that has poor access to the brain, was ineffective when given intravenously. Apparently the effect of propranolol was not due to the lowered threshold of mammary gland sensitivity to oxytocin which can be provided by beta antagonists. Slow infusions of adrenaline and isoprenaline failed to inhibit the milk ejection reflex. The alpha adrenergic antagonists, phentolamine and phenoxybenzamine, produced a dose-dependent inhibition of milk ejection reflex. Phentolamine in a dose of 1 mg/kg intravenously, increased milk ejection interval six-fold, while 2 mg/kg increased it thirteen-fold. This inhibitory effect of phentolamine and phenoxybenzamine was not due to the secondary hypotension induced by these drugs since hexamethonium could induce hypotension without affecting milk ejection. Picrotoxin and hyoscine, centrally active drugs, had no effect on the milk ejection reflex. Thus, the adrenergic mechanism involved in the secretion of oxytocin to suckling-induced milk ejection was slightly different from osmoreceptor of oxytocin in dehydrated states. In the milk ejection reflex, alpha receptors appeared to be excitatory while beta receptors were inhibitory. In the dehydrated animal, alpha receptor blockade as well as alpha and beta receptor stimulation were stimulatory to oxytocin release.

In contrast, Moos and Richards (231) concluded that beta nonadrenergic receptors are inhibitory to the release of oxytocin. Isoproterenol, a beta noradrenergic agonist, inhibited milk ejection in rats, but the response could be overcome by administration of exogenous oxytocin. Thus, the inhibitory effect was due to a central nervous blockade of oxytocin release. Propranolol, a beta noradrenergic antagonist, alone had no effect on the milk ejection response, but obtunded the inhibitory effect of isoproterenol on milk ejection in the lactating rat. Thus, it was suggested that the suckling-induced oxytocin release in lactating rats was facilitated via the alpha receptors and inhibited via the beta receptors.

Cholinergic Mechanism

Cholinergic mechanisms are probably no less involved in the secretion of oxytocin than adrenergic mechanisms. Clarke et al. (70, 71, 74) concluded that the neural pathway controlling the reflex release of oxytocin during suckling in the rat probably involved a cholinergic component. In urethane-anesthetized lactating rats, the suckling-induced milk ejection reflex could be inhibited by the nicotinic cholinoceptor antagonists mecamylamine and hexamethonium in a dose-dependent manner (70). In contrast, the muscarinic antagonists, atropine, hyoscine and benzhexol, failed to block the suckling-induced release of oxytocin even in high doses. Intraventricular injection of the cholinomimetics, acetylcholine, bethanecol and carbachol, stimulated a sustained release of oxytocin which could be abolished by atropine but not by mecamylamine and hexamethomium. It is interesting that none of the antagonists used affected either the release of oxytocin secondary to electrical stimulation of the neurohypophysis or the mammary sensitivity to oxytocin. Thus, a cholinergic component acting through the nicotinic receptors was probably involved in the neural pathway controlling the suckling-induced reflex release of oxytocin in the rat. While a second cholinergic component involving the muscarinic receptors might also exist, the evidence was less clear. These observations on the cholinergic mechanisms have been extended to include correlations with electrocorticograms recorded from electrodes placed over the frontal lobes of the lactating rats (73). During the milk ejection reflex, electrocorticogram activity showed a continuous slow-wave pattern compared to large amplitude slow waves alternating with low amplitude fast waves at rest. Carbachol and bethanecol given intraventricularly, inhibited milk ejection and increased the electrocorticogram activity. Atropine promoted milk ejection and an electrocorticogram with a slow-wave pattern. Beta adrenergic receptor antagonists failed to change the electrocorticogram pattern irrespective of any effect on milk ejection.

Opioid Peptides

Opioid peptide modulation of the effect of dopamine on oxytocin release using the isolated rat posterior pituitary gland provided additional information (372). Higher amounts of dopamine, which is the predominant catecholamine throughout the pituitary gland, were found in the posterior lobe than in the anterior lobe. The release of oxytocin from the isolated posterior lobe of the pituitary of untreated rats and rats pretreated with α-methyl-p-tyrosine (α-mPT) were studied. Under resting conditions, ouabain which released transmitter, was able to induce a much higher amount of oxytocin release from preparations pretreated with α-mPT. Dopamine failed to affect oxytocin release in resting tissues from control rats, but significantly reduced oxytocin secretion in tissues from dopamine-deficient rats. In contrast, the opioid peptide, β-endorphin or D-Ala^{2}-Pro^{5}-enkephalin amide, enhanced oxytocin release from untreated animals but not from animals pretreated with α-mPT. The effect of the opioid peptides could be blocked by naloxone. Thus it appeared that the release of oxytocin from the neurohypophysis may be inhibited by pituitary dopamine but enhanced by an endogenous opioid peptide. The opioid peptides probably exerted their effect through a disinhibitory phenomenon by removing the inhibitory effect of dopamine on oxytocin release, possibly through inhibition of dopamine release.

With a slightly different approach, using urethane-anesthetized lactating rats, it was shown that opiates inhibited the electrically induced release of oxytocin by an action on or close to the axon terminals within the neurohypophysis (75). The electrical activity of oxytocinergic neurons of urethane-anesthetized lactating rats was studied and the release of oxytocin was determined by measuring intramammary pressure. All oxytocinergic neurons studied displayed explosive 2 - 4 second bursts of electrical activity about 10 seconds before each milk ejection. Morphine given intravenously (2 mg) did not modify the electrical activity but oxytocin was not released. Intravenous administration of naloxone restored the release of oxytocin associated with the bursts of electrical activity. Intraventricular carbachol produced oxytocin release which was reduced by pretreatment with morphine and restored by naloxone. Oxytocin released by electrical stimulation of the neurohypophysis was substantially reduced within 6 - 12 minutes of giving 2 - 4 mg of morphine but was restored by naloxone (1 mg/kg). The intramammary pressure response to intravenous oxytocin was neither attenuated by morphine nor potentiated by naloxone. These findings suggested that morphine reduced the amount of oxytocin released but did not change the electrical activity produced by the oxytocinergic cells. Therefore, the action of morphine must be on or close to the terminals within the neurohypophysis because the oxytocinergic cells were in the hypothalamic nuclei. Since identical bursts of electrical activity occurred after treatment with morphine and naloxone, it is probable that no opioid synapse was directly involved in the transmission of the suckling stimulus from the nipples to the hypothalamus.

Administration of naloxone (10 mg/kg) intravenously to lactating rats increased suckling-induced rises in intramammary pressure without changing the freqency of milk ejection. Likewise, naloxone was also more effective in potentiating the effect of electrical stimulation of the oxytocinergic cells and in the amount of oxytocin release induced by carbachol. These observations suggested either that urethane could have a sustained opiate-like action, reversed by naloxone or, more probably, that endogenous opioid peptides tonically suppressed the release of oxytocin. These authors have some preliminary evidence that β-endorphin and (D-Ala, Met)-enkephalin could similarly inhibit oxytocin release.

The ability of morphine and its analogs, butorphanol and oxilorphan, to inhibit oxytocin release during lactation in mice was also demonstrated using milk yield as the index (162). When these substances were given intraperitoneally to lactating mice, the milk yield was significantly reduced, but a considerable amount of milk was obtained if oxytocin was then injected into the animals. The total milk yield remained unchanged since the residual milk yield was greater after treatment with morphine or its analogs.

Ionic Control

Previous evidence indicated that the release of neurohormones and neurotransmitters was dependent upon the concentration of calcium ions, probably at the internal face of the neurosecretory membranes. In order to characterize the role of the sodium channel in the release process of neurohypophyseal nerve terminals, Nordmann and Dyball (250) investigated the effects of veratridine on hormone release and calcium movements in the rat neurohypophysis *in vitro.* Veratridine is known to depolarise excitable cells by holding the sodium channels in an open state. Veratridine was found to increase hormone release dramatically, but the release rate was not sustained and declined to 90% after two hours. Veratridine increased Ca^{2+} uptake into the isolated neurohypophysis. This could be prevented by the addition of tetrodotoxin or the calcium antagonist D600 to the medium. Efflux of Ca^{2+} was, however, not affected by the addition of veratridine. Hormone release was prevented by the removal of Na^{+}, and the addition of tetrodotoxin or D600 and Mn^{2+} ions. Thus it appears that both Na^{+} and Ca^{2+} channels are important in the regulation of neurohypophyseal hormone secretion at the neurocellular level. Increasing K^{+} in the medium by ten-fold stimulated oxytocin release by increasing Ca^{2+} uptake into the neurohypophyseal cells (116). When hormone release declined, it could be increased again by veratridine.

Prolonged stimulation of the neural lobe has been shown to raise K^{+} in the incubation medium and to cause a significiant decrease in oxytocin release, probably because of inactivation of the late calcium channels and not because of terminal depletion. Dyball and Shaw (115) showed that prolonged electrical stimulation of animal neural lobes reduced oxytocin release significantly, but this decrease was abrogated if the Na^{+} in the incubation medium was reduced from 150 to 100 nM. Such a reduction in oxytocin release was not due to a failure of the action potentials generated by the stimulus since cyclical stimulation for half an hour followed by half an hour rest did result in a rise of oxytocin release back to the original level. Their results suggested that oxytocin secretion can be inactivated during prolonged electrical stimulation *in vivo*. The inactivation was more pronounced when extracellular Na^{+} was reduced from 150 to 100 mM and probably involved the late calcium channels.

Since exposure of the preoptic-tuberoinfundibular system to iron ions caused ovulation in the proestrus rat, Dyer (118) determined the effect of electrochemical deposition of iron on rat oxytocinergic neurons. In contrast to electrical stimulation of more than 15 Hz invariably inducing the release of oxytocin with milk ejection in the rat, electrochemical deposition of iron into the oxytocinergic neurons of the paraventricular-neurohypophyseal tract failed to release oxytocin.

Osmotic Pressure

The effects of osmotic stimulation on the response of the oxytocinergic and vasopressinergic neurons in the supraoptic nucleus were studied by characterizing their electrophysiologic response (46, 47). Using urethane-anesthetized rats, oxytocinergic cells of the supraoptic nucleus were identified by their characteristic demonstration of a burst of activity before reflex milk ejection. Those cells without response to milk ejection were considered vasopressinergic neurons. Oxytocin cells responded to a hypertonic injection of sodium chloride with a smooth, sustained three-fold increase in firing rate, but did not respond to isotonic sodium chloride pressure. Both oxytocin and vasopressin cells had reduced firing rates when plasma osmotic pressure was reduced by 10 - 15 mOsmol/kg. Although both oxytocin and vasopressin cells were osmo-responsive, phasic firing was characteristic of stimulated vasopressinergic neurons. It was therefore suggested that osmotic stimulation might allow discrimination between oxytocin- and vasopressin-secreting neurons. This study was expanded to include oxytocin release by measuring circulating plasma oxytocin by RIA following osmotic stimulation of oxytocinergic neurons of the paraventricular and supraoptic nuclei (48). Using the same animal model and method for identifying oxytocinergic and vasopressinergic neurons electro-physiologically, plasma oxytocin concentration was found to increase from 2.1 ± 0.3 μU/ml in the control period to 10.9 ± 2.8 μU/ml thirty minutes after an intraperitoneal injection of 1 ml of 1.5 M sodium chloride and 14.8 ± 2.8 μU/ml after a second injection of the hypertonic sodium chloride. Oxytocin-ergic neurons responded to an increase in plasma osmotic pressure after injections of hypertonic solutions of lithium hydrochloride or mannitol in a similar manner as when plasma osmotic pressure was raised by sodium chloride. Thus these studies demonstrated that oxytocin cells in the neurohypophyseal system responded to the osmotic pressure of the blood plasma rather than to sodium or chloride concentration, with an accompanying increase in circulating plasma oxytocin levels.

Water deprivation in the lactating rats resulting in chronic dehydration activated both vasopressin and oxytocin neurons but there was a marked difference in the pattern of their response (375). Water deprivation for 0 - 24 hours in lactating rats increased the firing rate of oxytocinergic neurons from 0.9 spikes/sec to 2.8 spikes/sec after 24 hours. Unlike vasopressinergic neurons, which showed an increased number (84 - 100%) of cells firing phasically after dehydration, only 4% of the oxytocin cells showed phasic firing which changed from a slow, irregular to a fast, continuous discharge after water deprivation. In the dog, both hemorrhage and hypertonic saline stimulated the release of oxytocin and arginine vasopressin (393). Plasma oxytocin increased significantly after phlebotomy and also after infusion of hypertonic saline which elevated plasma osmolality and sodium significantly. The log plasma oxytocin was highly correlated with the volume of blood removed. Hemorrhage had a greater stimulation on arginine vasopressin whereas the stimulation of both hormones was comparable after hypertonic saline. In contrast, fetal hemorrhage did not stimulate fetal oxytocin release in the pig (213).

Prostaglandins

Several findings, in both women and animals, have indicated that $PGF_{2\alpha}$ could cause oxytocin to be released; the postulated mechanism of release of the $PGF_{2\alpha}$ was on the hypothalamo-neurohypophyseal system. However, other studies could not obtain a release of oxytocin when $PGF_{2\alpha}$ was given intraperitoneally in lactating rats. Two recent studies examined the effect of $PGF_{2\alpha}$ on the oxytocin release in the rat (5) and the pig (122, 123). The effect of $PGF_{2\alpha}$ given into the third cerebral ventricle on oxytocin and vasopressin release was monitored by recording the firing activity of the oxytocinergic and vasopressinergic fibers (5). The firing activity of paraventricular nucleus neurons was recorded extracellularly in urethane-anesthetized rats and 50 neurons were identified as neurosecretory neurons. Thirty neurons were identified as oxytocinergic because they gave a burst of action potential 12 - 15 seconds before milk ejection and the remaining 20 paraventricular neurons were classified as nonoxytocinergic since no response was found prior to the milk ejection reflex. An increased firing rate followed the intraventricular injection of $PGF_{2\alpha}$ (500 mg in 1 ml isotonic saline) in 83% of the 30 oxytocinergic neurons and lasted for 20 - 30 minutes. In contrast, 75% of the 20 nonoxytocinergic neurons showed decreased firing activity in response to intraventricular $PGF_{2\alpha}$. No change was seen in any of the neurons studied with intraventricular isotonic saline. With $PGF_{2\alpha}$, the intramammary pressure was also slightly increased. Thus it appeared that in the rat, during lactation, intraventricular injection of $PGF_{2\alpha}$ had a differential effect on the paraventricular neurons, stimulating the oxytocinergic neurons but suppressing the nonoxytocinergic neurons.

Ellendorf et al. (123) determined if prostaglandins could alter plasma oxytocin levels at times other than just before parturition and whether the response was consistent with a positive or negative feedback. A single intramuscular injection of 5 mg of prostaglandin $F_{2\alpha}$ induced a significant increase in plasma oxytocin levels, from 1.16 ± 0.25 μU/ml to 13.37 ± 1.80 μU/ml, 15 60 minutes after the injection in sows six days postpartum, and from 0.76 ± 0.17 μU/ml to 9.69 ± 2.20 μU/ml during estrus. Vasopressin levels did not show any significant change. They concluded that circulatory steroid levels did not interfere with $PGF_{2\alpha}$-induced oxytocin release.

Estrogen

Estrogen stimulated the release of oxytocin in the female rat and exerted its effects mainly by activating the anterior neural input to the neurosecretory cells in the hypothalamo-neurohypophyseal system (406). These investigators were, however, unable to demonstrate unequivocally any effect of estrogen on the posterior neural input to the neurosecretory cells. The effect of estrogen treatment (estradiol benzoate, 20 mg/day for two days) on plasma oxytocin determined by bioassay, and vasopressin determined by bioassay and RIA was investigated in ovariectomized rats under urethane-anesthesia (1 - 2 g/kg). Plasma oxytocin concentrations in the jugular venous blood of ovariectomized, estrogen-treated rats (46.2 - 232.2 μU/ml, median = 105.5 μU/ml) were significantly higher than those of ovariectomized control rats (0 - 130 μU/ml, median = 46.0 μU/ml) ($p \leqslant 0.05$). Plasma vasopressin did not show any significant change with or without estrogen treatment. The facilitatory effect of estrogen on oxytocin release in rats was affected by anterior (hypothalamic) deafferentation but not by posterior deafferentation. Thus, the effect of estrogen appeared to be one of activating the anterior neural output to the neurosecretory cells of the hypothalamo-neurohypophyseal system.

Thyrotropin-Releasing Hormone

Thyrotropin-releasing hormone (TRH) stimulated the release of both oxytocin and vasopressin in rabbits *in vivo* (394). TRH, when given in a dose of 250 nmol/kg, had no significant effect on oxytocin release, but a significant increase in plasma oxytocin and vasopressin was seen with a five-fold increase in the TRH dose. In contrast, intraventricular injection of as little as 3nM of TRH produced a significant release of oxytocin and vasopressin, a release which occurred much earlier than with the intravenous route. The active TRH analog, MK-771, but not the inactive analog, D-tyrosine$_2$ TRH, had similar effects on oxytocin and vasopressin release as the TRH did.

SECRETORY DIRECTION

Oxytocin and vasopressin were detectable in the cerebrospinal fluid, probably a reflection of direct secretion from the hypothalamic neurons in the third ventricle (108). In intact and hypophysectomized rats, basal plasma oxtocin levels were similar to the increased levels seen after anesthesia. Oxytocin release into the blood was partially impaired after hypophysectomy since, unlike the intact animal, cerebrospinal fluid withdrawal did not result in any further increase of plasma oxytocin. However, cerebrospinal fluid levels of oxytocin remained unchanged after hypophysectomy. In contrast, hypophysectomy greatly reduced plasma arginine vasopressin, which became detectable only after removal of cerebrospinal fluid, but elevated arginine vasopressin levels in the cerebrospinal fluid. It was suggested that oxytocin fibers could be more rapidly regenerated than arginine vasopressin fibers from the hypothalamus to the neurohypophysis, and the compensatory increase in hypothalamic vasopressinergic neurons might account for the elevated cerebrospinal fluid vasopressin levels. The biologic significance of the oxytocin and vasopressin in the cerebrospinal fluid remain to be established, but it might be a route by which the hormones exert their neurobehavioral effects.

■ ■ ■

Chapter 3

▪ EFFECT ON THE CENTRAL NERVOUS SYSTEM

Neurohypophyseal hormones have been found to have profound effects on the learning and memory processes. In this respect the effect of vasopressin is relatively better understood than that of oxytocin. Many investigations have been undertaken to clarify the neurobehavioral and memory effects of oxytocin in relation to vasopressin. The effect of oxytocin on anterior pituitary hormone secretion has been sketchy, and the information available on its effect on gonadtropin is inconclusive. Based on rather preliminary findings, oxytocin and vasopressin have been suggested to have an influence on opiate analgesic and alcohol tolerance. These effects of oxytocin on the central nervous system are discussed below.

NEUROBEHAVIORAL AND MEMORY EFFECTS

The supraoptic-neurohypophyseal system has been known to be involved in the acquisition and maintenance of adaptive behavior. de Wied et al. (101, 102) reviewed the role of neurohypophyseal hormones and behavior. Much of the focus was on vasopressin, but it will form the background to some of the recent reports on oxytocin behavior. It should also be noted that many of the investigations were carried out in animals, specifically the rat. Removal of the posterior pituitary lobe attenuated shuttle-box avoidance in rats. It appeared that vasopressin and related peptides were involved in the learning and memory processes and in sexually-motivated approach behavior but not in maintenance of a food running approach response in hungry rats. Secondary to failure to produce vasopressin, rats of the Brattleboro strain which have hereditary hypothalamic diabetes insipidus, were inferior in acquiring a two-way, active-avoidance reflex and had severe memory loss that could be restored by vasopressin administration. Vasopressin is effective both in pharmacologic and physiologic doses with respect to these functions. The investigations on neurohypophyseal hormones, learning behavior and memory were usually performed using animal models (the rat) which were subjected to classical or instrumental conditioning of either a passive- or active-avoidance behavior response. The site of the behavior effects of the neurohypophyseal hormones and related peptides appeared to be in the posterior thalamic area and the parafascicular nuclei, but destruction of the latter did not totally abrogate such effects of systemically administered vasopressin. The learning processes appeared to reside in the rostral septal and dorsal hippocampal areas which, if extensively destroyed, also appeared to prevent the behavior effect of vasopressin. Arginine vasopressin was the most potent peptide on learning behavior, followed by lysine vasopressin, pressinamide, proly-argyl-glycinamide, and oxytocin in that order. Arginine vasotocin was less potent than oxytocin with reference to learning behavior. Removal of the C terminal glycinamide of vasopressin decreased the potency by 50%. There appeared to be a dissociation between the endocrine and behavioral activities of neurohypophyseal hormones. Through the effect of a membrane-bound hypothalamic exopeptidase, oxytocin was converted to prolyl-leucyl-glycinamide (PLG) which inhibited the release of melanocyte-stimulating hormone (MSH), potentiated the behavioral effect of L-DOPA, and protected against puromycin-induced amnesia in mice. Oxytocin might have a behavioral effect opposite to that of vasopressin.

There is some controversy as to whether oxytocin facilitated extinction of pole-jumping avoidance behavior and if it did, it was only 20% as potent as arginine vasopressin. It was thought that this effect of oxytocin resulted from counteracting the behavioral effects of vasopressin and was therefore most marked in dehydrated rats but not in normal nondehydrated rats. Oxytocin facilitated physical dependence on morphine; the C terminal part of the hormone is necessary for this effect. Puromycin induced amnesia was reversed by oxytocin and related peptides that facilitate physical dependence on narcotic analgesics.

Several publications appeared on the effect of oxytocin and vasopressin on learning and memory, and some of the possible mechnisms by which these peptides acted. Oxytocin reinstated maternal olfactory cues for nipple orientation and attachment in rat pups (321). When the female rat's ventrum was washed with alcohol and acetone to remove milk from the nipple area, there was a marked deficiency in nipple orientation and attachment in her 8 - 9-day-old pups. This nipple attachment was restored, even when the nipples were ligated, if the thoroughly-washed mother was injected with oxytocin. Atropine blocked this effect of oxytocin, but acetylcholine restored it. Rat and bovine milk had no effect in restoring the nipple orientation. It was suggested that maternal release of the putative substance for the olfactory cue for nipple orientation was oxytocin-dependent and probably involved cholinergic receptors. This substance was presumably excreted from the aprocrine Montgomery glands around the nipple. These glands are surrounded by myoepithelial cells that contract in response to oxytocin. Admittedly, the dose of oxytocin used (10 U/kg) was greater than endogenous levels of oxytocin released at milk ejection.

Using the step-down passive avoidance behavior to study the effect of neurohypophyseal hormones on rat behavior, it was found that oxytocin shortened the step-down latency considerably but that lysine-8-vasopressin lengthened it (192). However, neither oxytocin nor vasopressin had any effect on the step-on latency. The dose of oxytocin and vasopressin used was 300 mU/kg for each peptide, given intraperitoneally.

Bohus et al. (38) extended their earlier observations that oxytocin was an amnesic neuropeptide. This property of oxytocin and vasopressin was shown to act through the *consolidation* and *retrieval* processes. Experiments performed on rats indicated that intracerebroventricular administration of 1 mU of oxytocin resulted in the attenuation of the passive avoidance response taught to these animals in a time gradient manner.

Such attenuation was strongest when oxytocin was given 0 to 23 hours after the learning trial, and at least six hours after the learning trial, suggesting that there was a significant time interval-dependent effect on the process of *consolidation* of acquired experience. The shorter the interval between the learning and treatment, the more pronounced was the amnesia effect. Therefore, oxytocin can be considered to be an amnesic neuropeptide, modulating the consolidation process of the learning experience. Since oxytocin, when administered one hour prior to the retention test, also attenuated the passive avoidance behavior, it was concluded that the amnesic action of oxytocin was also mediated via the *retrieval* process. Intraventricular administration of arginine vasopressin (AVP) resulted in the facilitation of the passive avoidance behavior with optimal effect seen when AVP was given 0 to 23 hours after the learning, and minimal effect seen when given at three hours. Thus, AVP *enhanced* the consolidation and retrieval components of the learning experience, while oxytocin *attenuated* both processes and acted as an amnesic neuropeptide. Similarly, oxytocin antiserum and AVP antiserum produced the opposite effects of the respective neuropeptides on the memory processes.

The site of action and the effect on catecholamine changes when oxytocin and vasopressin were injected into limbic-midbrain structures to produce memory consolidation in the rat were determined (191). Oxytocin injected bilaterally into the hippocampal dentate gyrus (25 pg/side) or into the midbrain dorsal raphe nucleus (50 pg) signficantly *attenuated* passive avoidance behavior in the rat. However, when oxytocin was injected into the dorsal septal nucleus, *facilitation* of passive avoidance behavior resulted. By contrast, vasopressin, when injected into all three nuclei, facilitated passive avoidance behavior. Both oxytocin and vasopressin had no effect on memory consolidation when injected into the central amygdaloid nucleus. Injection of vasopressin into the hippocampal dentate gyrus increased the disappearance of norepinephrine (NE) in the dentate gyrus and in the nucleus ruber; injection into the dorsal septate nucleus decreased NE disappearance in this nuclei but increased it in the nucleus ruber; injection into the dorsal raphe nucleus increased the disappearance of dopamine in the locus coeruleus and in the nucleus ruber. Thus, memory consolidation can be oppositely influenced by minute amounts of oxytocin or vasopressin given locally into certain limbic-midbrain structures; suggesting that these areas of the brain are involved in the memory effects of these peptides. While modulation of catecholamine turnover in specific brain areas might be related to this behavioral effect after vasopressin administration, unpublished data from these authors indicated that oxytocin failed to modulate the disappearance rate of NE and dopamine in the limbic-midbrain areas. Thus, it is unlikely that the opposite effects of oxytocin and vasopressin on memory consolidation involve NE and dopamine.

Telegdy and Kovac (352) confirmed the antagonistic effect of oxytocin to vasopressin on learning and extended it to the effect of these neurohypophyseal hormones on brain neurotransmitter metabolism. Oxytocin (300 mU/kg intraperitoneally) decreased the norepinephrine (NE) contents of the hypothalamus, septum and striatum significantly within ten minutes without affecting the dopamine (DA) contents of these regions. Oxytocin increased the 5-hydroxytryptamine (5HT) in the hypothalamus but decreased it in the septum. The DA turnover in the hypothalamus was increased but both NE and DA turnovers in the striatum were decreased. Oxytocin was ineffective on both the acquisition of an active-avoidance reflex and on the step-on latency of passive-avoidance reflex, but shortened the step-down latency. Only with a large dose of oxytocin (300 mU/kg) was the extinction of the active-avoidance reflex facilitated. When these findings of oxytocin were compared to those obtained with vasopressin, it was clear that the brain neurotransmitter metabolism was different following oxytocin or vasopressin treatment. It appeared that oxytocin increased DA turnover in the hypothalamus but decreased the turnover of both catecholamines in the striatum.

In contrast to many of the investigations employing a conditioned or learned behavior response, Delaney et al. (98) examined the behavioral effects of neurohypophyseal hormones and peptides in mice that had not been previously conditioned. When vasopressin, lysine vasopressin (LVP), oxytocin and arginine vasotocin were given in doses of 0.1 - 1.0 mg into the cerebral ventricles, a dose-dependent behavioral response was obtained with all the peptides. The response included marked hyperactivity, extensive foraging, increased grooming and, at higher doses, stereotyped scratching, squeaking and barell-rolling. All four hormones were equipotent, which was strikingly different from the opposite effects of oxytocin and vasopressin on the response to a learned behavior. Thus the effect observed in this study appeared to be a spontaneous, direct behavioral effect as opposed to the effect on the memory processes involved in learning.

Loss of the N-terminal group as with DDAVP, or the C-terminal group as with desglycinamide lysine vasopressin, significantly reduced the potency in eliciting the behavioral changes reported. This effect of the neurohypophyseal hormones was not due to vasoconstriction of effects on the choroid plexus since pretreatment with ergotamine, $PGF_{2\alpha}$ and ethoxolamide given intraventricularly did not alter the LVP-induced behaviors. Pretreatment with reserpine, haloperidol and physostigmine given intraperitoneally, attenuated the locomotion and grooming but not the behavior changes. Pretreatment with α-methyl-p-tyrosine, p-chorophenylalanine, 6-hydroxydopamine and diphenhydramine, given either intraperitoneally or intraventricularly, did not have any effect on the behavior changes. None of these drugs, D-amphetamine or nicotine mimicked the behavior changes, and neither could LVP. It is likely that the behavior effects of neurohypophyseal hormones were modulated through an effect on the brain, but it is noteworthy that the doses of oxytocin and vasopressin were rather large.

Maternal Behavior

Oxytocin was shown to be capable of inducing maternal behavior in virgin rats (266). When injected into the cerebral ventricles, oxytocin in a dose of 0.4 mg was able to induce full maternal behavior towards foster pups in 42% of the virgin female rats, compared to vasopressin and saline which had no such effect. The maternal behavior identified included 1) grouping the foster pups together in a corner; 2) licking the pups; 3) crouching in such a manner as to accomodate the pups beneath the foster mother; 4) nest building and; 5) retrieving the pup if it was removed from the rest of the group. Such behavioral changes continued for as long as ten days if the pups were allowed to stay with the test animal. Elevated levels of estrogen were necessary for the induction of full maternal behavior by oxytocin since most of the virgin rats which responded were in the last day of diestrus, in proestrus or estrus. Ovariectomized virgin rats treated with estradiol benzoate were able to display full maternal behavior after oxytocin treatment, whereas ovariectomized animals not primed with estrogen did not respond to oxytocin.

THE PITUITARY GLAND

There have been suggestions that the posterior pituitary hormones can influence the release of anterior pituitary hormones. There is also a very rich vascular portal system in the pituitary stalk which can drain hypothalamo-neurohypophyseal secretions. Previous reports of high concentrations of vasopressin in the hypophyseal portal circulation of the monkey pituitary and its role in regulating corticotropin release certainly confirm that there is an endocrine communication between the adenohypophysis and the neurohypophysis. Several investigators have therefore examined the effects of oxytocin or its agonists on the release of pituitary hormones.

Prolactin

The effect of oxytocin on prolactin production and release by rat pituitary cells (2B8) was compared to other neurohypophyseal peptides and pineal hormones (164). Both oxytocin and arginine vasopressin had no effect on prolactin secretion from the rat pituitary cells, but physiological concentrations of arginine vasotocin and the pineal hormone, melatonin, stimulated prolactin secretion from rat pituitary cells.

Similarly, *in vitro* administration of oxytocin and lysine vasopressin failed to stimulate prolactin release in humans (99). Oxytocin given intravenously in doses of 1 U and orally in doses of 50 U three times a day for seven days to normal adult females and to postpartum women did not produce any significant change in their plasma prolactin over a 90-minute period during which frequent blood samples were obtained. The amount of milk produced in the postpartum women was not modified. Similarly, 5 or 10 U of lysine vasopressin given intramuscularly did not change the plasma prolactin levels in spite of a significant reduction in excretion of free water. Thus, both vasopressin and oxytocin given systemically in the doses cited above did not affect prolactin release.

Gonadotropin

The effect of neurohypophyseal hormones on pituitary gonadotropin release has been somewhat unclear because of the conflicting results so far reported. The administration of oxytocin to human adults either produced elevations of follicle-stimulating hormone (FSH) and luteinizing hormone (LH) separately or in combination or caused no change in the gonadotropins. Part of the problem was due to the purity of the oxytocin and part of it was due to the imperfections of the methods for measuring gonadotropins. Dawood et al. (97) infused oxytocin intravenously into male adults in stepwise incremental doses up to a maintenance dose of 256 mU/min and separately up to 132 mU/min to maintain a steady state. With circulating blood levels of 378 pg/ml and 228 pg/ ml during the steady state, no significant change in circulating FSH and LH levels was found. Thus, it appeared that even in large pharmacologic doses given systemically, oxytocin did not have any significant effect on gonadotropin release.

Using the *in vitro* pituitary perfusion technique, Caffrey et al. (60) demonstrated that oxytocin in concentrations of 0.02, 0.2 and 20 mU/ml stimulated LH release and increased the baseline secretion of LH significantly. However, 2 mU/ml of OT did not.

When added to gonadotropin releasing hormone (GnRH)-containing perfusion media, 0.02 mU oxytocin/ml produced significant enhancement and prolongation of the LH response to GnRH, but higher and lower concentrations of oxytocin did not. Oxytocin in a dose of 0.02 mU/ml was able to convert previously ineffective doses of arginine vasopressin (2 and 20 mU/ml) into effective doses for enhancing GnRH-induced LH release. While these findings indicated that both oxytocin and vasopressin might have a role in reproduction via an interaction with GnRH on the anterior pituitary, *in vivo* findings in some species did not indicate any significant change in gonadotropin levels even with pharmacologic doses of oxytocin.

The Effect on Melanocyte Stimulating Hormone

Neurohypophyseal hormone fragments including 1) analogs of H-Cys-Tyr-Ile-Gln-Asn-OH with oxytocin-like sequences shortened stepwise from the NH_2-terminus; 2) an identical series of peptides but terminating at the carboxyl terminus in an asparagine amide moiety rather than asparagine and; 3) a selected number of analogous peptides, with a vasopressin-like sequence instead of the oxytocin-like sequence, were found to stimulate release of melanocyte stimulating hormone (MSH) (63). On a molar basis all the peptides possessed the same activity except for H-Gln-Asn-OH which was inactive. Serum MSH increased while pituitary MSH decreased.

DRUG DEPENDENCE/TOLERANCE

The role of oxytocin and vasopressin as endogenous mediators of opiate analgesia or tolerance/dependence was thought to be minimal at best (299), which was in contrast to an earlier suggestion. In mice, both oxytocin and vasopressin failed to demonstrate any alteration of chronic morphine effects and pretreatment with either neurohypophyseal hormone failed to alter brain concentrations of acutely injected morphine.

Neurohypophyseal hormones were shown to influence ethanol dependence and tolerance in mice. Arginine vasopressin had a significantly more potent effect that oxytocin in maintaining ethanol tolerance, an observation which was similar to their relative potencies in maintaining the recall of learned responses (171).

CEREBRAL BLOOD FLOW

Unlike previous observations that neurohypophyseal hormones affected peripheral blood flow and blood pressure, Lassoff and Altura (198) reported that direct, local application of synthetic, preservative-free oxytocin and vasopressin to the pia of the anesthetized rat failed to exert any influence on either microvascular diameter or arteriolar blood flow of the brain.

■ ■ ■

Chapter 4

▪ PHYSIOLOGIC EFFECT IN NONPREGNANCY

THE UTERUS

Oxytocin-Prostaglandin Relationship

Previous studies have shown that oxytocin promoted the release of prostaglandin from the rat uterus. Such a relationship between oxytocin and prostaglandin continued to be explored under slightly different conditions. The effects of oxytocin on prostaglandin release from the rat uterus under different levels of estrogen were investigated (4). Prostaglandin release was decreased by estradiol but increased by oxytocin if estradiol was given six hours earlier, but not if given 24 hours earlier. This suggested that estradiol caused an *accumulation* of prostaglandin while oxytocin caused a *release* of it. However, with a longer interval after estradiol treatment, prostaglandin-degrading enzymes in the uterus were stimulated and thus prostaglandins were metabolized and were then available for release. The release of prostaglandin from the uterus of ovariectomized rats treated with estradiol for six hours and given oxytocin, was inhibited by indomethacin, an inhibitor of prostaglandin biosynthesis. The prostaglandin releasing action of oxytocin was completely abolished by treatment with phospholipase C which specifically catalyzed hydrolytic cleavage of phospholipids to diglyceride and corresponding phosphoryl products. This reduced the availability of arachidonic acid, the rate-limiting precursor in the biosynthesis of prostaglandins.

Several studies showed that the effects of oxytocin on uterine contraction were not mediated through the release of $PGF_{2\alpha}$. In isolated rat uteri, specific oxytocin antagonists such as [1-L-penicillamine] oxytocin had no effect on uterine response to prostaglandin but completely suppressed the response to oxytocin, while prostaglandin synthetase inhibitors such as indomethacin and diclofenac sodium had no effect on the uterine response to both prostaglandin and oxytocin (65). Because the parturient uterus is sensitive to oxytocin action, one postulate was that this sensitization was related to the enhanced prostaglandin synthesis and release induced by oxytocin. This hypothesis was tested in the rat uteri (65). Uterine prostaglandin was low in nonpregnant uteri and non-pregnant estrogen-treated uteri, but increased progressively during pregnancy to more than 100% on the day of expected delivery. Oxytocin induced contractions in both pregnant and nonpregnant uteri with a further *release* of prostaglandins in the former but not in the latter. Prostaglandin *content*, however, was not significantly changed. Therefore it appeared that the marked increase in sensitivity of the uterus to oxytocin was related to the enhanced prostaglandin synthesis and release induced by oxytocin. Rat uterus produced maximum $PGF_{2\alpha}$ accumulation into the incubation medium *in vitro* during estrus. Oxytocin in doses of up to 500 mU/ml did not affect the $PGF_{2\alpha}$ accumulation at any stage of the estrus cycle (177). This suggested that the uterotonic action of oxytocin was independent of prostaglandin participation.

In order to determine if $PGF_{2\alpha}$ released from the uterus by oxytocin mediated the oxytocic action of oxytocin, the changes in uterine motility and rate of secretion of $PGF_{2\alpha}$ at different stages of the estrus cycle of sheep were recorded while oxytocin was infused directly into the uterine artery (282). The uterine response and $PGF_{2\alpha}$ secretion varied with the phase of the estrus cycle, but neither increased synthesis of $PGF_{2\alpha}$, an essential intermediate step in the activation of the myometrium by oxytocin, nor increased contractile activity were necessary for the oxytocin-induced synthesis of $PGF_{2\alpha}$.

When the effects of oxytocin and vasopressin on uterine activity and their interactions with prostaglandins were compared, oxytocin was found to stimulate uterine activity in pregnant and nonpregnant rabbits, while vasopressin was stimulatory in the nonpregnant, weakly stimulatory in late pregnancy, and inhibitory during the first three days of pregnancy (199). Indomethacin reduced the responses to both oxytocin and vasopressin in the nonpregnant animals but had no effect on the pregnant rabbits. $PGF_{2\alpha}$, PGE, or PGE_2 given to 6 - 8-day pregnant rabbits inhibited the stimulatory effects of vasopressin slightly but had no effect on the uterotonic action of oxytocin. It was suggested that both neurohypophyseal hormones could, under certain hormonal conditions, stimulate the uterus by mechanisms in which prostaglandins were not required.

The mechanism of action of bradykinin and oxytocin and their relationship to $PGF_{2\alpha}$ on the rat uterus were studied (395). Acetylcholine and $PGF_{2\alpha}$ produced concentration-effect curves on the rat myometrial preparation similar to that seen on the whole uterus. However, bradykinin and oxytocin produced a shift to the right with a decrease in maximum response on myometrial preparations compared to those observed on the whole uterus. Indomethacin depressed the action of all four agonists, but had a greater suppression on the action of bradykinin and oxytocin. The effects of bradykinin and oxytocin on myometrial contraction were greatly enhanced by prior sensitization by $PGF_{2\alpha}$. It was concluded that the mechanism of action of bradykinin and oxytocin on the rat uterus involved both a direct action and an indirect one, the latter possibly involving the release of prostaglandins from the endometrium.

Whalley and Raynes (396) extended this study to determine if prostaglandins were involved in the action of oxytocin on the rat uterus *in vitro* and *in vivo*. When applied to the myometrial surface of uterine horn preparation from rat in estrus, oxytocin (0.25 - 2 mU) induced contractions in a dose-dependent response, but ten times the dose (2.5 - 20 mU) was required to induce a weak response when applied on the endometrial surface. Indomethacin had no effect on responses when applied to the myometrial surface. In contrast to this, oxytocin (0.1 - 1 mU/ml) added to the tissue bath for 20 - 30 minutes on the myometrial, but not the endometrial surface produced prolonged cyclic activity that could be attenuated by indomethacin (5 mg/ml) applied to either uterine surface. Likewise in the conscious estrus rat, oxytocin (0.05 - 2 mU/kg) and indomethacin (2 mg/kg) produced similar results. Therefore, it was concluded that oxytocin had a significant, direct effect on myometrial activity and that a prostaglandin component affecting mainly the cyclic uterine activity was involved.

Thus, from these recent reports it is still unclear, even in the rat, whether the action of oxytocin involves a release of prostaglandin from the uterus of the nonpregnant animal. There may well be species differences in this mechanism of oxytocin action. Some component of the effect of oxytocin on the myometrium may require or involve prostaglandin release and the importance of this component may vary from species to species and may be influenced by ovarian hormones.

The Influence of Sex Steroids

It is becoming increasingly clear that the action of oxytocin and prostaglandin on the nonpregnant myometrium can be sensitized by the level of sex steroid hormones. Recordings of the electrical activity of the rat myometrium revealed that PGF_2 and oxytocin induced myometrial excitability with a marked increase in sensitivity during the last stage of gestation and a 1000-fold increase at parturition compared to midpregnancy (195). At midpregnancy, the sensitivity to both PGE_2 and oxytocin was lower than in the nonpregnant uterus. Estradiol treatment increased the sensitivity to both PGE_2 and oxytocin, but less than that seen during parturition, while progesterone treatment reduced the sensitivity to below that of castrated rats. Analysis of the data indicates that the sensitivity of the myometrium to PGE_2 and oxytocin during early and midpregnancy could be stimulated by progesterone but that the sensitivity during late gestation and parturition could be mimicked by estrogen and progesterone.

The sensitivity of the myometrium of the sow was increased under the influence of estrogen both *in vitro* and *in vivo*. With 25 mU oxytocin/ml, all myometrial strips taken from sows in estrus responded *in vitro* with an initial increase in baseline tone and a subsequent increase in the rate of contraction (81). The duration of response varied from 15 to 40 minutes. When oxytocin (1 - 5 U/animal) was given to the sow during estrus *in vivo*, there was initially a small increase in uterine tone followed by an increase in the frequency of contraction from 6 contractions per 10 minutes to 8 - 10 contractions per 10 minutes. The response usually lasted 10 minutes.

The effect of oxytocin on myometrium *in vitro* has mainly been studied using animal uteri. The spontaneous activity of *human* myometrial strips obtained during the proliferative and secretory phases of the menstrual cycle and during the early and late stages of pregnancy had recently been reported (144). In Munsick's solution (a balanced physiological medium), human myometrium usually relaxed rapidly, became maximally active in 30 minutes and remained active for up to 28 hours. Spontaneous myometrial activity was seen in both nonpregnant myometria and pregnant myometria from different stages of gestation. Indomethacin, in doses of 1 μmol in the bathing fluid, was able to abolish the uterine activity for up to two hours while oxytocin (10 - 30 mU) induced myometrial activity in less active myometria (both pregnant and nonpregnant), but had no superimposed effect on tissue with high-grade spontaneous activity. $PGF_{2\alpha}$ was more potent to both pregnant and nonpregnant tissue and increased the resting tone if the tissues showed high-grade spontaneous activity. The inhibitory effect of indomethacin was overcome by $PGF_{2\alpha}$ but not by oxytocin. It was hypothesized that oxytocin probably exerted its effect on human myometria through or in conjunction with endogenous prostaglandin. This might help to explain the safety of intravenous oxytocin "titration" regimes and the excessive uterine activity that might result from exogenous prostaglandin administration. Nevertheless, we know clinically that, given inappropriately, oxytocin can induce excessive uterine activity.

Oxytocin and Other Hormones

Vasoactive intestinal polypeptide (VIP), a highly basic octacosapeptide is found in neurons, besides the gut, and represents a neurotransmitter. VIP-containing nerve fibers innervate small vessels and smooth muscle cells, are vasodilatory and relax smooth muscle. The hypothesis that VIP could play a role in the nervous regulation of smooth muscle contraction was examined (255). VIP did not have any effect on the nonoxytocin-stimulated myometrium grafted in a transparent ear chamber, but produced a dose-dependent inhibitory effect on oxytocin-stimulated myometrium. There was no effect on the oxytocin-stimulated myometrium with a VIP dose of 6 pmol/min/kg, a slight inhibition at 25 pmol/min/kg, and complete inhibition at 100 pmol/min/kg. Since VIP nerve fibers and terminals were present in the myometrium, it was postulated that VIP played a physiologic role as a neurotransmitter in the oxytocin-induced activity of the myometrium. Oxytocin, PGE_1, PGE_2 and $PGF_{2\alpha}$ can overcome the uterine relaxant effect of relaxin (64). Rat uterus treated *in vitro* with relaxin reduced the amplitude of contractions but when subsequently given 250 mU of oxytocin, showed a significant increase in contraction amplitude compared to controls.

Albasri and Walker (6) were able to show that there was a compensatory increase in oxytocin secretion in the Brattleboro rat, a species which cannot make vasopressin. Urinary excretion of oxytocin in this rat was 6.7 ± 1.3 mU/24 hours, while in the normal, male long Evans rat it was 0.58 ± 0.083 mU/24 hours.

Oxytocin Antagonism

Tetrahydroisoquinoline salsolinol was able to significantly antagonize the responses of isolated smooth muscle tissues to oxytocin and vasopressin but not to bradykinin (163). The salsolinol appeared to act as a competitive antagonist for oxytocin on its uterine receptor. There was some suggestion that salsolinol might be formed *in vivo* after alcohol consumption. If this is true, then salsolinol formation after ethanol consumption may have physiological significance, and may explain a possible mechanism by which ethanol inhibits uterine contraction.

Biochemical Changes

The effects of oxytocin on the biochemical changes induced at a cellular level in the uterus have provided some understanding of the mechanism of oxytocin-induced myometrial contractions. The effects of oxytocin on cyclic nucleotide levels of estrogen-primed rabbit myometria revealed that oxytocin stimulated, while methacholine inhibited adenylate cyclase activity, but both stimulated guanylate cyclase activity in the myometrium (298). In the absence of phosphodiesterase inhibitor 1-methyl-3isobutylxanthine (MIX), methacholine increased guanosine 3′, 5′-cyclic monophosphate (cGMP) levels but had no effect on adenosine 3′, 5′-cyclic monophosphate (cAMP) levels. Oxytocin increased cAMP but had no effect on cGMP levels. MIX alone increased both cAMP and cGMP levels and the addition of oxytocin or methcholine further increased cGMP levels, indicating activation of guanylate cyclase. The activation of guanylate cyclase by oxytocin, but not by methacholine, was abolished in Ca^{2+}-free solution. Oxytocin also increased cAMP levels above those produced by MIX, but metacholine decreased them. None of the effects seen were altered by indomethacin – indicating that prostaglandins did not modulate them. The findings of this study did not support a role for cGMP in either the initiation or maintenance of myometrial contraction, and indicated that oxytocin stimulation of adenylate cyclase was not mediated by prostaglandins.

The myometrial contractile effect of oxytocin was mediated, in part if not wholly, by the movement of calcium across the sarcoplasmic membrane which preceded the contractile act of the smooth muscle. Increased temperature and adenosine triphosphate (ATP) enhanced the accumulation of calcium in the microsomal fraction resembling sarcoplasmic reticulum isolated from uterine muscle (62). If this microsomal fraction was preloaded with Ca^{2+}, accumulation occurred under the effect of ATP and reached a steady state of bound calcium when the ATP was exhausted. Oxytocin, PGE_2, $PGF_{2\alpha}$, and the antibiotic ionophores, X537A and A23187, released such previously accumulated calcium. Only oxytocin released the intrinsic calcium. Wanner et al. (381) employed a computer simulated model to determine the likely mechanism by which calcium was involved in oxytocin-induced myometrial contractions. Oxytocin intervention in the calcium distribution within the myoplasm was mimicked by a discontinuous change of the rate constants and compared with an experimental time profile of uterine tension. The possible mechanisms included calcium release from the cell membrane, acceleration of calcium transport from the extracellular space by a "gate" mechanism, release from intracellular organelles, or blockade of the calcium pump; all of which could lead to an increase in myoplasmic calcium concentration. When the computer models were screened, they showed that oxytocin must act predominantly via the release of calcium bound to the cell membrane. However, comparisons of such simulated models did not take into consideration the kinetics of oxytocin distribution in the myometrium. Thus, comparisons of time profiles do have limitations.

Purified microsomal fraction from pregnant and nonpregnant human uteri bound calcium in the presence of ATP; the binding increased with advancing gestation (61). Progesterone increased, while oxytocin decreased such ATP-dependent calcium binding to the microsomes. Progesterone combined with oxytocin counteracted each other. The action of progesterone was specific since biologically inactive progesterone analogs were ineffective on calcium binding. The reduction of ATP-dependent calcium binding to microsomes resulted in increased unbound calcium and was thus consistent with the *in vivo* contracting action of oxytocin. The opposite held true for progesterone.

THE CERVIX

In estrogen-treated ovariectomized rats, uterine luminal fluid accumulation had been used as an indirect measure of cervical tone, with loss of fluid indicating reduced tone. The mechanism by which analogs of PGE_2 and $PGF_{2\alpha}$ caused a loss of luminal fluid was probably secondary to myometrial contractions and not due to reduced cervical tone. Kennedy (185) found that oxytocin did not inhibit uterine fluid formation but reduced uterine luminal fluid in a dose-dependent relationship with a rapid effect seen 15 minutes after injections with supramaximal doses of oxytocin. The effect of oxytocin on luminal fluid accumulation was not affected by prostaglandin biosynthesis and indomethacin, but loss of the fluid occurred slightly sooner in oxytocin-injected controls than in the indomethacin-treated groups. The dose-response relationship between oxytocin and uterine luminal fluid was reduced by indomethacin. While these findings indicated that prostaglandins acted synergistically with oxytocin to cause luminal fluid loss, the validity of using luminal fluid accumulation as an index of cervical tone remained open to question.

THE MENSTRUAL CYCLE

It is unclear what oxytocin does during the menstrual cycle. Plasma oxytocin was studied in the menstrual cycle of eight rhesus monkeys at 3 - 4 day intervals. Oxytocin concentrations were significantly higher in mid-cycle (days 10 - 11) than at days 4 - 5 and days 28 - 29 (126). There was no relationship between oxytocin and estrogen or progesterone. These findings are somewhat difficult to reconcile since the means of the concentration of oxytocin ranged from 30 to 60 pg/ml, which were considerably higher than those reported, by the same authors, for maternal levels in pregnancy in the rhesus monkey (288). If the data is carefully examined, it does appear that the peak oxytocin level was on days 8 - 9. This tends to be in agreement with other observations that estogens stimulate the release of oxytocin and neurophysin II, the latter having been shown to be elevated in mid-cycle in the rhesus monkey.

In the rat, pituitary levels of oxytocin and vasopressin fluctuated at different phases of the estrus cycle and norepinephrine might have been mediating some of these changes (84). Pituitary levels of oxytocin and vasopressin were maximal on the morning of proestrus, declined during estrus, and were lowest at metestrus. Norepinephrine levels in the paraventricular nucleus were decreased on proestrus and estrus, compared to metestus-diestrus. Concentrations of norepinephrine in the supraoptic nucleus did not vary or change in ovariectomized rats treated with either estrogen or progesterone or both ovarian hormones. In contrast, estradiol benzoate administered to ovariectomized rats induced the elevation of pituitary oxytocin content which was not increased by further treatment with estrogen or progesterone. Ovarian hormone treatments decreased steady-state levels of norepinephrine in the paraventricular nucleus and this could be abrogated by the catecholamine synthesis inhibitor α-methyltyrosine. This suggested that, during the estrus cycle, the ovarian hormones modulated pituitary neurohypophyseal hormone levels and therefore release through norepinephrine turnover in the paraventricular nucleus.

THE OVARY

The effect of oxytocin on the ovary has been studied extensively. Oxytocin was able to induce ovarian contractions. Recent investigations have focused on the role of oxytocin in ovulation and in the corpus luteum. Roca et al. (285) found significantly reduced ovulation in rabbits, induced by human chorionic gonadtropin (hCG), when a specific antiserum to oxytocin was administered to the rabbits. Indomethacin, by blocking $PGF_{2\alpha}$ synthesis, was also able to inhibit hCG-induced ovulation in the rabbit, but in a dose-dependent manner. Significant inhibition by indomethacin was only observed at a dose of 7.5 mg/kg, or more. They concluded that oxytocin played a physiological role in ovulation in the rabbit. The mechanism of oxytocin action in ovulation was probably through the stimulation of the smooth muscle of the ovary, and therefore ovarian contraction followed by follicular rupture. While combined treatment with oxytocin antiserum and indomethacin had a greater inhibitory effect on ovulation, the results could not establish if there was a relationship between oxytocin and prostaglandin during ovulation. In contrast, sheep that were actively immunized against oxytocin conjugated to bovine serum albumin and emulsified in Freund's complete adjuvant (1 mg/animal four times at weekly intervals) had prolongation of the luteal phase of estrus without any evidence of ovulation inhibition (131, 314). In five ewes immunized with oxytocin bovine serum albumin conjugate, the length of the cycle was significantly prolonged by 3.7 ± 1.5 days (mean ± S.E.M.) compared to cycle lengths of 15 - 19 days before immunization and to control animals immunized with bovine serum albumin only. There was a significant correlation between prolongation of the luteal phase and oxytocin antibody concentration in the individual animal. The progesterone levels showed that the life span of the corpus luteum was prolonged. A high oxytocin antibody concentration was also accompanied by interference with estrus behavior. The effect of the oxytocin antibody on luteal prolongation did not appear to be due to the central nervous system effect since the cerebrospinal fluid had very low antibody titres. Oxytocin concentrations increased significantly with immunization against oxytocin and could be directly correlated with the circulating antibody concentration. The authors hypothesized that the luteolytic action of estradiol-17B in the sheep might be mediated through a stimulatory effect on endometrial oxytocin receptor concentration. Since the release of $PGF_{2\alpha}$ was thought to be potentiated by both estradioal and oxytocin, it was suggested that the luteolytic action of estradiol and oxytocin was through an effect on uterine $PGF_{2\alpha}$ production. It has previously been proposed that a rise in follicular estrogen induced an increase in endometrial oxytocin receptors, thereby facilitating the stimulatory effect of oxytocin on endometrial $PGF_{2\alpha}$ production.

In heifers, the intramuscular administration of 100 U of oxytocin was accompanied by a rise in plasma PGF (240). The increase in posterior vena cava plasma PGF was significantly influenced by the day of the estrus cycle, with the greatest response seen three days after estrus. The influence of estrogen was demonstrated in ovariectomized heifers where the level of plasma PGF increased 24 hours after priming with estradiol. No further increase was observed after oxytocin was injected. Peak levels of PGF were higher in the plasma of the posterior vena cava than in the jugular vein, thus suggesting the source of the PGF to be in the pelvic area. These findings provided some explanation for earlier observations that the daily administration of oxytocin to heifers after the estrus markedly shortened the ensuing diestrus and the overall length of the estrus cycle. In heifers which had been given oxytocin for seven days beginning on day 1and which had been hysterectomized on day 2, the corpus luteum was maintained, suggesting that a uterine luteolysin was necessary. The apparent luteolytic effect of oxytocin was due to its ability to stimulate a rise in circulating levels of PGF which were probably from the uterus.

The suggestion that oxytocin could be luteolytic through a release of $PGF_{2\alpha}$ was used as the basis for examining the effect of exogenous oxytocin on luteal function in mares. Daily injections of 150 U of oxytocin on days 5, 6, 7 and 8 after ovulation failed to induce luteolysis, and normal plasma progesterone concentration and ovulatory intervals were maintained (236). Pretreatment of the animals with estrogens 24 hours before oxytocin injection on day 7 after ovulation, also had no luteolytic effect.

Oxytocin was able to induce ovarian contractility in the rat both *in vitro* and *in vivo* (286, 287). *In vitro* ovarian contractility in response to oxytocin (0.5 - 25 mU/ml) remained the same throughout the estrus cycle of the rat (287). However, experiments carried out *in vivo* showed that a greater response was observed on the afternoon of proestrus (286). It had been postulated that when the progesterone levels increased, a withdrawal of estrogens from the receptor sites allowed a reduction of intracellular cAMP. The lower concentration of intracellular cAMP might therefore result in an enhanced oxytocin effect. The differences between the responses *in vivo* and *in vitro* could be due to the denervation and vascular interruptions in the latter situation.

■ ■ ■

Chapter 5

■ PHYSIOLOGICAL EFFECTS DURING PREGNANCY

The role of oxytocin in parturition remains controversial and continues to attract attention. Previous studies have been limited in number and many used bioassay, which entails problems. Whether circulating oxytocin levels increased during pregnancy and whether there was any change in oxytocin levels during parturition remains unclear. What is probably more widely accepted is the increasing sensitivity of the human uterus with gestation. It is also clear that oxytocin must play some role in uterine activity at parturition since it was successfully used in inducing uterine activity at term. The view that the fetus has a role in determining the onset of parturition was becoming more popular and fetal oxytocin release has been suggested as an important consideration. Finally, the role of oxytocin in milk ejection, which has been more widely accepted than the role of oxytocin in parturition, was further examined. Thus investigations were focused on oxytocin during pregnancy, parturition and lactation in several species, and also on the interaction between oxytocin and prostaglandin in the pregnant uterus.

UTERINE SENSITIVITY TO OXYTOCIN

It is well established that the sensitivity of the uterus to oxytocin increases as gestation advances towards term. This principle is being applied to possible predictive use and therapeutic manipulation. In a retrospective analysis, the sensitivity of the uterus to oxytocin as found during the oxytocin challenge test was analyzed to examine whether it could identify patients delivering pre-, at or post-term (347). Baseline uterine contractility was similar in patients who went into pre-term, term and post-term delivery. Both baseline contractility and oxytocin sensitivity increased in all three groups of patients. However, uterine sensitivity to oxytocin increased significiantly in the pre-term group and decreased in the post-term group compared to the term group. While the authors suggested that the uterine response to oxytocin could be used to identify patients who would deliver pre- or post-term, the study had its drawbacks. The authors admit to many confounding variables which govern the accuracy of the external tocodynamometer. Such an oxytocin sensitivity test on a sensitive pre-term uterus may send a patient into irreversible pre-term labor.

In a group of 108 women with prolonged pregnancy, the administration of estrogen to some of these women increased the oxytocin sensitivity of the uterus in 33%, but ATP given either alone or in combination with estrogens and vitamin B_1 increased the uterine sensitivity in 75% (275). When estrogen alone, for up to 12 hours, failed to enhance myometrial sensitivity as determined by the oxytocin sensitive test of Smyth, ATP was able to convert such uteri to a sensitive state. Thus ATP resulted in a quicker and greater sensitization of the myometrium to oxytocin than estrogens.

An apparently good correlation was found between uterine activity and plasma oxytocin between 40 and 120 pg/ml when oxytocin was given in mid-trimester pregnancy, but there was no correlation at all higher plasma levels (369). The lack of response at lower plasma oxytocin levels at this stage of pregnancy is not surprising since it is generally accepted that the uterus is relatively insensitive to oxytocin at this stage of gestation compared to term. The lack of correlation with uterine activity at higher plasma levels of oxytocin could be due to their RIA method, which had been criticized (392), and not solely due to other factors that were implicated.

OXYTOCIN AND THE PREGNANT UTERUS

In vitro studies on the role of prostaglandin in spontaneous and oxytocin-induced contractions of the 18 to 21-day pregnant rat uteri showed that while spontaneous contractions *in vitro* were directly related to prostaglandin production, oxytocin-induced contractions did not depend on increased prostaglandin production (114). On the contrary, uterine contractions could be induced by oxytocin even when there was a simultaneous reduction in prostaglandin production as determined by RIA, although a minimum threshold of prostaglandin might be required for maximum response to oxytocin.

A new approach to the relationship between oxytocin and prostaglandins was reported when it was found that prostacyclin production might play an important role in modulating the action of oxytocin. Prostacyclin is produced by the endothelial cells of the blood vessels and has a vasodilatory effect. Thus far, prostacyclin has been shown to relax the uterus. The release of prostacyclin from chopped myometrial fractions of 18 to 20-day pregnant rats were measured and the effect of oxytocin, $PGF_{2\alpha}$ and bradykinin were examined (400). Pre-incubation of the myometrial tissue with 10 mU/ml of oxytocin increased prostacyclin production from 2.25 ± 0.48 to 3.75 ± 0.73 ng/mg over a period of 15 minutes. Bradykinin also cause a significant increase in myometrial prostacyclin but 1 μg/ml of $PGF_{2\alpha}$ did not signficiantly increase prostacyclin release. An interesting observation was that there was a much greater increase in prostaglandin output in the presence of oxytocin in uteri at 18 - 20 days pregnancy than in uteri on day 22 (400), an observation similar to that made for the effect of oxytocin on $PGF_{2\alpha}$ (65). It was explained that at 18 - 23 days gestation, basal arachidonic acid metabolism was low and oxytocin stimulated phospholipase A_2 enzyme activity, whereas at day 22 basal synthesis of prostaglandins was much higher and thus there was little capacity for further synthesis even with additional stimulation by oxytocin.

Fructose-1,6-dipholphate can abolish ocytocin-induced spastic inertia in rat uterus (82). This finding has led to the use of fructose-1,6-diphosphate, 5 g in 50 ml of water, as an intravenous infusion in 30 deliveries having oxytocin induction of labor. The findings showed a significant reduction in the recovery of uterine contractions following uterine inertia compared to controls (82).

In sheep pretreated with oxytocin, betamimetics such as ritodrine, bunitrolol and oxytocin offset the oxytocin action and improved utero-placental blood flow (202).

MEASUREMENTS OF OXYTOCIN

The RIA of oxytocin is gaining popularity since the sensitivity and specificity of the assay confer advantages over bioassays which require larger volumes of blood, are tedious and have biologic variability. Nevertheless, the use of RIA continued to be confined to a few laboratories because of the difficulty of raising good antiserum and because of technical difficulties with the RIA. Several investigators reported the development of radioimmunoassays for oxytocin in their laboratories with and without extraction of the plasma samples, and with variable extent of validation of their assays (91, 92, 97, 151, 155, 280, 297, 368).

Dawood et al. (92) and Raghavan et al. (280) used the same antibody that they had produced, and employed extraction of their samples with fuller's earth. The recoveries were in the order of 60% from the fuller's earth, or 40% from the starting biological sample (plasma or amniotic fluid), which were similar to what was obtained by Chard and his colleagues who also used fuller's earth. The oxytocin was subsequently eluted from the fuller's earth with acetone. The labeled oxytocin was iodinated using the lactoperoxidase enzyme technique by Dawood et al. (92), but by the chloramine-T method and omission of the sodium metabisulfite by most of the other investigators (151, 155, 280, 296). The use of an enzymatic labeling procedure kept damage to the hormone to a minimum during the procedure. Other investigators (151, 296) also employed acetone as their extractant. However, some workers continued to use unextracted plasma for their RIA (155, 368) and the levels of oxytocin obtained by some investigators (368) have been consistently much higher than results from any of the RIA methods of oxytocin that have been described. The problems of using unextracted plasma in the RIA of oxytocin have been discussed previously and were referred to again by several workers (90, 92, 93, 393). Nonspecific interference with binding by plasma proteins, hormone degradation by enzymes such as peptidase in the plasma, and poor validation of some of the assays made it difficult to reconcile the widely differing results obtained.

In general, the better validated assays indicated good specificity of the antiserum with less than 1 - 5% cross-reaction with arginine vasopressin, vasotocin and some oxytocin and vasopressin analogs (92, 151, 280). The sensitivity of the assays ranged from 1 pg to 5 pg (2 pg = 1 μU), and most reports indicated satisfactory precision. In one laboratory, it was demonstrated that intravenous infusion of oxytocin produced plasma oxytocin levels that correlated significantly with the dose of oxytocin (92) which was a further confirmation of the precision and accuracy. Correlation of the RIA with bioassay of oxytocin was demonstrated in only two of the studies (92, 280).

Robinson and Walker (284) described the use of agarose-bound neurophysin for the extraction of oxytocin from biological fluids such as plasma, urine and cerebrospinal fluid. They had recovery rates of 96 to 98%. The main advantage of this extraction method appeared to be its specificity. Columns could be used to extract and concentrate dilute samples or to help identify small amounts of neurohypophyseal hormones by affinity chromatography. Although this extraction method could prove useful for RIA, immunoreactive fragments which retain their ability to bind to agarose-neurophysin may be present. Finally, for the routine RIA of large numbers of samples, it might prove both expensive and difficult to obtain neurophysin.

PREGNANCY

Previous studies by bioassay and indirect estimations have indicated that maternal plasma oxytocin levels increased during pregnancy. The few studies using RIA gave divergent conclusions. Several investigations have recently been undertaken to determine maternal oxytocin levels during pregnancy in women (92, 93, 94, 96), rhesus monkeys (228), baboons (9), and miniature pigs (351).

In one study using RIA that did not employ preliminary extraction of the plasma samples, high levels of oxytocin, from 130 to 326 pg/ml, were reported in the plasma during pregnancy at 24 - 42 weeks (368). These levels were rather high compared to most published reports where RIA or bioassay was employed for the measurement of oxytocin. The difficulties of the RIA of oxytocin using unextracted plasma or serum samples have been discussed many times by different authors. Nonspecific inhibition of binding between oxytocin and oxytocin antibody by plasma constituents and the possibility of plasma enzymes degrading both radioactive and nonradioactive ligands are only two of the many problems that have been pointed out. The lack of extraction of the plasma samples and the failure to validate adequately the RIA method employed were also pointed out (392).

Dawood et al. (94, 96) used a specific and sensitive RIA of oxytocin with preliminary extraction of the plasma samples with fuller's earth in 362 samples obtained throughout pregnancy. Mean maternal plasma oxytocin increased from 17.9 pg/ml at 7 weeks gestation to 21.3 pg/ml at 40 weeks gestation. There was a significant correlation between maternal plasma oxytocin and gestational age. A significant rise in maternal plasma oxytocin from 26.4 ± 4.8 pg/ml at 38 weeks gestation to 74.2 ± 14.2 pg/ml at 39 weeks gestation followed by a significant drop to 21.3 ± 5.3 pg/ml at 40 weeks gestation was noted but the significance of such a rise remained unexplained. More than 80% of the maternal plasma samples had detectable oxytocin levels.

Human amniotic fluid had detectable oxytocin levels in 90% of the samples measured (94, 96). Oxytocin levels in amniotic fluid increased from 7.8 ± 4.6 pg/ml at 14 - 15 weeks gestation to 43.9 ± 14.7 pg/ml at 40 weeks, and decreased to 30.8 ± 10.5 pg/ml at 41 - 42 weeks. The main source of oxytocin in the amniotic fluid was thought to be the fetal urine, since it was shown to have good concentrations of oxytocin (95).

Preliminary results, based on serial maternal plasma oxytocin levels throughout pregnancy in ten women, suggested that *poor* secretion of oxytocin as opposed to *good* secretion of oxytocin was associated with dysfunctional labor due to abnormal uterine activity which was not due to fetopelvic disproportions (96). Poor secretion of oxytocin was defined as levels of plasma oxytocin which were persistently less than 10 pg/ml during the third trimester of pregnancy.

One study (200) was however, unable to show any increase in plasma oxytocin in pregnant women compared to nonpregnant women, which was in striking contrast to many of the studies that have shown such an increase during pregnancy. Further, estrogens have been demonstrated to stimulate the release of oxytocin (Chapter 2).

In pregnant baboons, 71 of 75 maternal plasma samples (95%) during pregnancy (110 - 180 days gestation) had detectable oxytocin levels measured by RIA (91). There was a highly signficant correlation between maternal plasma oxytocin and gestational age. Maternal plasma oxytocin levels were usually between 10 and 80 pg/ml. When oxytocin was given through a continuous intravenous infusion pump into the baboon at 171 - 179 days gestation, there was a significant correlation between the maternal plasma oxytocin achieved and uterine activity recorded with an open-ended intrauterine catheter. Measurements of oxytocin in the umbilical vein, and jugular and cardiac blood from the fetus, and in the maternal uterine vein and maternal peripheral blood showed that in late pregnancy the fetal circulation had a higher concentration than the maternal compartment.

In the rhesus monkey, maternal plasma oxytocin was detectable in 16 of 42 samples while amniotic fluid was detectable in 20 of 37 samples. The mean maternal plasma oxytocin ranged from 20.7 pg/ml to 27.4 pg/ml between 100 and 169 days gestation (228). There was, therefore, no significant change in maternal plasma oxytocin in the last half of pregnancy. Careful examination of the standard error of the mean showed that there were obvious wide differences in the individual plasma oxytocin values. Mean amniotic fluid oxytocin levels ranged from 10.9 to 13.5 pg/ml between 100 and 169 days gestation with no significant change in gestational age. Amniotic fluid always had a lower oxytocin concentration than the maternal plasma. The authors therefore concluded that in the rhesus monkey there was no significant change or trend in maternal plasma oxytocin and amniotic fluid oxytocin in the latter half of pregnancy.

Thus, recent investigations in pregnant women, rhesus monkeys and baboons showed that circulating maternal oxytocin levels were increased compared to nonpregnancy levels. The significance of increased oxytocin levels is unclear, but increased maternal estrogen levels are probably responsible for the increase in plasma oxytocin.

Taverne et al. (351) measured oxytocin levels in the miniature pig in late pregnancy and correlated it with myometrial electrical activity and plasma progesterone and estrogens. During late pregnancy, there was no change in plasma steroids and plasma oxytocin remained below 1.2 μU/ml (2.4 pg/ml) 12 to 2 days before parturition. Although a significiant fall in progesteronc and a significant increase of estrogens were detected at -24 to -10 hours before expulsion of the first piglet, myometrial activity, as recorded by electromyographs, reflecting regular electrical discharges accompanied by elevations in plasma oxytocin levels (4.0 - 8.0 μU/ml) (8.0 - 16.0 pg/ml) appeared at -9 to -4 hours before this first birth.Therefore, in spite of no significant change in maternal plasma oxytocin in the miniature pig, the evolution of myometrial activity for up to nine hours before delivery and until completion of delivery coincided with increasing maternal plasma oxytocin concentrations.

Oxytocin appeared to play an important role in the control of monoaminergic function during late pregnancy (260). The metabolic transformation of epinephrine to metanephrine and acid metabolites was increased in peripheral organs such as the liver, spleen, ovary and uterus, and in the brain after hypophysectomy or adrenalectomy in 19 - 21-day pregnant rats. Oxytocin pretreatment (1.25 U/100 g) on days 20 and 21 decreased the metabolic transformation of epinephrine in most organs.

LABOR

The role of the oxytocin reflex during coitus, birth and breastfeeding were summarized in a presentation at a symposium, but most of the references cited were either based on indirect measurements of oxytocin or on empirical observations (241).

Dawood et al. (90, 92, 93, 94, 139) have extensively studied the levels of plasma oxytocin during spontaneous labor and parturition. Serial measurements in 11 patients throughout the first, second and third stages of labor showed that there was a significant increase in plasma oxytocin from the first to the second stages of labor, followed by a significant decline from the second to the third stages. During the first stage of labor, the mean maternal plasma oxytocin level was 43.1 ± 11.2 pg/ml, which was similar to the mean levels found near term (92, 93). During the second stage of labor, the mean plasma oxytocin was 114.1 ± 20.7 pg/ml, which declined to 63.9 ± 10.8 pg/ml in the third stage. Thus, oxytocin was present in the maternal circulation during the first stage of labor, and minute to minute plasma oxytocin levels did not correlate with any specific phase of uterine contractions. The levels were only slightly higher than during late pregnancy.

Similar observations on maternal plasma oxytocin increase during delivery of the fetus have been made by other investigators, although the absolute levels of oxytocin reported differed quite considerably, from 12.0 pg/ml in one study (200) to as high as 210 - 460 pg/ml in another (368). Difficulties with the latter study have been discussed above.

In the miniature pig, measurements of plasma oxytocin from the jugular vein by RIA revealed that during the initial stage of labor, there was a small increase of oxytocin of up to 24 μU/ml, accompanied by a concomitant increase in frequency of uterine contraction as recorded by electromyography using bipolar electrodes implanted in the uterus (134, 350, 351). Plasma oxytocin further increased during the expulsive phase of labor, reaching a peak concentration of 70 μU/ml at the moment of delivery. At this point uterine contractions occurred at the rate of 9 - 15 contractions per hour and lasted between 1.5 and 3.5 minutes. There was another rise in plasma oxytocin with the expulsion of the placenta, when the frequency of uterine contractions reached its highest rate (15 - 20 per hour) and lasted 40 - 80 seconds. The surge in maternal plasma oxytocin was remarkably similar to that seen in women during the second stage of labor. Thus in the miniature pig, a rise of plasma oxytocin was first seen, accompanied by an evolution of uterine activity, beginning about nine hours before delivery of the piglet, with further surges in maternal plasma oxytocin seen at the expulsion of the piglets. A peak of oxytocin (65 pg/ml), coinciding with expulsion of the calf, was also reported in cows during parturition (296).

Indirect observations supporting the role of oxytocin in rat parturition were recently summarized (206). Phase-like increases in the frequency and amplitude of uterine contraction during the onset of labor could be the result of pulsatile release of oxytocin. 1 to 2 mU of oxytocin increased the contractile activity of the sensitized rat uterus at term for 10 - 15 minutes. The rat uterus did not respond until 3 - 6 hours before delivery and from that time the sensitivity increased exponentially. These observations suggested an important role of oxytocin in rat parturition, but no satisfactory demonstration of a rise in oxytocin has been reported.

Vaginal distension in the goat significantly elevated jugular vein plasma oxytocin in one pregant and two nonpregnant goats (27). This reflex was attenuated by pretreatment with dexamethasone phosphate, 8 mg daily for four days in nonpregnant goats. Progesterone infusion was not always inhibitory to this reflex. Thus, the authors suggested that the inhibition of oxytocin release might play a role in the placental retention associated with corticoid-induced deliveries.

Lumbar anesthesia could effectively block the Ferguson reflex. In untreated sheep, vaginal distension induced a rise in jugular venous oxytocin and utero-ovarian venous PGF levels, but after lumbar anesthesia neither response occurred (130). However, lumbar anesthesia had no effect on the rise in utero-ovarian venous PGF level induced by the administration of oxytocin. Thus, the findings suggested that a reflex release of oxytocin was involved in the elevation of the utero-ovarian venous PGF observed after vaginal distension in the sheep.

Amniotomy in women did not produce any significant change in plasma oxytocin but increased plasma 13,14-dihydro,15-keto-prostaglandin $F_{2\alpha}$ (PGFM) (309). Therefore, the rise in PGFM after amniotomy was not through an effect of oxytocin.

FETAL OXYTOCIN

The role of the fetus in contributing oxytocin to the process of parturition has gained further support in studies in women and the rat.

Measurements of umbilical arterial and umbilical venous plasma oxytocin levels indicated that oxytocin levels were significantly higher in the umbilical artery than in the umbilical vein whether the patient was in labor or not, provided the mother was not receiving oxytocin (90, 94, 95). The umbilical arterial plasma oxytocin levels were significantly higher when the patients were in labor, 116 ± 17.2 pg/ml (spontaneous labor and vaginal delivery) and 118.4 ± 18.1 pg/ml (cesarean section after onset of labor), than when the patients were not in labor, 29.8 ± 7.5 pg/ml (cesarean section without labor). Likewise, the umbilical arteriovenous difference in plasma oxytocin was significantly higher in women after the onset of labor – indicating a flow of oxytocin from the fetal to the maternal side. The secretion rate of oxytocin from the fetal to the maternal compartment was estimated to be 5.5 to 6.0 ng (2.75 to 3.0 mU) oxytocin/min during spontaneous labor compared to 1.0 ng (0.5 mU)/min in the absence of labor. When the mother was given oxytocin, there was a higher plasma oxytocin level in the umbilical vein than in the umbilical artery, thus indicating that a gradient could not be produced in the opposite direction and that there was transplacental passage of oxytocin. Since the estimated secretion rates of oxytocin from the fetal side to the maternal compartments were similar in vaginal delivery and cesarean section after the onset of labor, this fetomaternal gradient was already established in the first stage of labor. Amniotic fluid oxytocin levels were higher in cesarean section with labor (48.1 ± 12.4 pg/ml) than cesarean section without labor (13.6 ± 6.7 pg/ml). Also, fetal urine had a mean oxytocin concentration of 53.8 ± 11.7 pg/ml. Thus, the data indicated that during the first stage of labor, the fetus was a significant source of oxytocin, infusing oxytocin towards the placenta and myometrium at a rate that was remarkably similar to intravenous infusion of physiologic doses of oxytocin for induction of labor. This was in keeping with the findings that during the first stage of labor, the maternal plasma oxytocin levels did not show any significant rise.

The role of fetal oxytocin in parturition gained support from studies in the rat. Administration of dopamine (150 mg/kg/10 hours), L-dopa (200 mg/kg/12 hours) and carbidopa (20 mg/kg/12 hours) to pregnant rats from the sixteenth day of gestation delayed delivery by 10 to 18 hours (301). Treatments with these three drugs inhibited the release of fetal oxytocin which occurred during spontaneous labor in the rat but did not affect the release of maternal oxytocin. The fetal pituitary gland content of oxytocin increased significantly with maternal treatment with these drugs as compared to control animals, but maternal pituitary gland oxytocin remained unchanged with or without treatment. Fetal brain tissues concentrated tritiated dopamine more than maternal tissues after tritiated dopamine was given to the pregnant rats. Thus it was very likely that fetal oxytocin was involved in the timing of parturition in the rat and that dopamine could postpone parturition by suppressing the release of fetal oxytocin.

LACTATION

Lincoln et al. (206) have summarized the evidence for oxytocin release in reflex milk ejection in rats. Nursing rats displayed an intermittent pattern of milk ejection during the prolonged period of nursing, characterized by uniform but transient rises in intramammary pressure at regular intervals of 5 - 15 minutes. This milk ejection reflex could be mimicked by a *bolus* injection of 0.5-1.0 mU oxytocin, which suggested that the natural process involved a pulsatile release of oxytocin. Electrical stimulation of the neurohypophysis or the hypothalamo-neurohypophyseal tract also produced the milk ejection reflex, the optimal stimulation occurring at 40 - 60 Hz for two to four seconds. Electrophysiologic recordings of the hypothalamic neurosecretory cells in rats demonstrated that the milk ejection was preceded by an explosive and synchronous acceleration in the firing of half the supraoptic and paraventricular neurons, with discharge rates increasing from 0 - 5 to 30 - 70 spikes/sec for a period of one to four seconds. The increase in the discharge rates was confined almost entirely to the continuously firing units.

The threshold for the suckling-induced reflex release of oxytocin is probably distinct and therefore different from the threshold for the release of prolactin (PRL) in some species. In the urethane-anesthetized rats, Wakerly et al. (376) showed that mothers which had previously suckled 12 pups demonstrated a graded increase in the amount of oxytocin released during a three hour suckling test when the number of pups suckling was increased from 6 to 8 to 10. In contrast, mothers which had suckled 6 pups during their lactation had maximum frequency of milk ejection whether 6, 8 or 10 pups were applied to the breasts. With 6 pups at the nipple, there was a significant rise in PRL levels with the second group of mothers but not the first. With 10 pups, PRL was released in both groups of mothers, but earlier and more so in the second group. Thus, the thresholds for suckling-induced oxytocin and PRL release were distinctly different and were in part governed by the preceding suckling experience of the mothers.

Hypotonicity can increase the sensitivity of oxytocin action on the lactating mouse mammary gland and this can therefore be employed to increase the selectivity of mammary gland bioassay for oxytocin but not its sensitivity (129). This explained the markedly enhanced response to oxytocin which had been previously shown when the Tyrode solution for the administration of oxytocin was diluted. The effect of hypotonicity on increasing the sensitivity of oxytocin was through an increased efficiency in conveying the effect of hormone receptor binding to the contractile apparatus.

In women, suckling during the first two to three days postpartum was associated with maternal secretion of oxytocin as determined by RIA (200). Maternal plasma oxytocin levels reached as high as 19.2 ± 3.4 pg/ml.

In cows, plasma oxytocin measured by RIA showed a marked response to the preparation of milking and a further effect was also noted during the attachment of the teat cups of the milking machine (296). Peak plasma oxytocin levels were 15 - 50 pg/ml. Similar findings were obtained in another study where plasma oxytocin reached 22 - 30 pg/ml, two minutes after milking in six cows (155).

A significant reduction of plasma oxytocin released in response to milking cows was found during estrus compared to after estrus, thus suggesting that ovarian sex steroid hormones influenced the release of oxytocin during milking (18).

The dose-response and time-response relationships between injected oxytocin and intramammary pressure were studied in normal lactating sows with the possibility that oxytocin could be employed for promoting lactation in veterinary practice (121). With 20, 40, or 80 U oxytocin given intramuscularly, there was a rapid initial increase in intramammary pressure to 15 to 27 mm Hg lasting for 22 - 32 seconds, followed by secondary oscillations in the pressure for 40 - 60 minutes. When frequency and amplitude of oscillations and total work were calculated, it was found that when compared to 20 U, 40 and 80 U of oxytocin produced an increased frequency of 150 and 249% and total work of 36.6 and 104.4% respectively, but no difference in amplitude. The authors therefore recommended giving larger doses of 50 - 80 U of oxytocin to obtain additional clinical advantage.

The same investigators measured the intramammary pressure response to oxytocin in healthy sows during the first eight weeks of lactation (318). During the second week of lactation, there was a 55% increase in response to oxytocin as compared with the first week but responsiveness declined from the second to the eighth week. Because the incidence of mastitis-metritis-agalactia in sows was high during the first week of lactation, the authors concluded that the low responsiveness of the mammary gland might be a contributory factor.

Adrenergic mechanisms are involved in the mammary response to oxytocin during lactation. Prostaglandins were demonstrated to inhibit the mammary response to oxytocin (373). In lactating rats, $PGF_{2\alpha}$ (4 and 8 μg/kg), PGE_1 and PGE_2 (2 and 4 μg/kg) given intravenously reduced the intramammary pressure response to intravenous oxytocin (300 μU) by 50 - 80%. While adrenergic blockade with phenoxybenzamine or propranolol did not affect the oxytocin-antagonist action of a single *intravenous* $PGF_{2\alpha}$ injection (4 μg/kg), alpha receptor blockade reduced the oxytocin-antagonist action of *infused* $PGF_{2\alpha}$ (8 μg/kg/min for 15 minutes) by 38%. However, beta or alpha and beta receptor blockade has no effect. In contrast, an α-receptor blockade eliminated or greatly reduced the oxytocin-antagonist effects of PGE_1 and PGE_2 given either as a single intravenous dose or as an intravenous infusion for 15 minutes. Beta adrenergic blockade has no influence and alpha and beta blockade reduced the oxytocin-antagonist effect of PGE_1 and PGE_2 by 70%. It was therefore concluded that the mechanism of PG-induced inhibition of the oxytocin response might involve mammary vasoconstriction and/or alterations in myoepithelial activity of cAMP and cGMP via adrenergic modulation.

The adrenergic influence on mammary response to stimulation was further examined in rats using a sightly different stimulus. Rhythmical, manual compression for 10 - 15 minutes of two thoracic mammary glands of urethane-anesthetized rats caused an immediate and significant reduction in intramammary pressure response in the abdominal mammary glands induced by intravenous oxytocin (158). This lasted for 13 minutes, followed by a return to normal response during which there was increased responsiveness significantly higher than before compression. Intravenous injection of 5 μg of epinephrine mimicked the response to manual compression, but pretreatment with the beta adrenergic blocker, propranolol, and pretreatment of the gland to be compressed with carbocaine had no effect. Manual compression of the mammary gland during the phase of recovery and increased responsiveness also had no effect. The findings suggested that the motor-inhibitory system of the rat mammary gland, which was composed of mainly sympathetic components, could be reflexly activated by mechanical stimulation of the mammary gland area and might account for the failure of rat pups to obtain sufficient milk during the first 10 - 15 minutes of suckling.

This work was further extended to show that electrical stimulation for 60 - 120 seconds of the cut end of a mammary nerve reduced the intramammary pressure response in the contralateral abdominal mammary gland to oxytocin by 50%. After stimulation the pressure response took 3 - 10 minutes to recover (225). The depressant effect of the nerve stimulation was overcome by bilateral adrenal ligature, intravenous injection of 0.3 mg of propranolol or transection of the spinal cord above the level of entry of the mammary nerve. The depressant effect of electrical stimulation of the mammary nerve on the intramammary pressure response to oxytocin could also be blocked by introducing milk intraductally to the gland whose pressure was being monitored. The authors concluded that afferent information from the mammary mechanoreceptors could reflexly antagonize the sympathetic adrenal system which had previously been activated.

■ ■ ■

Chapter 6

▪ PHARMACOLOGIC AND OTHER EFFECTS

PHARMACOLOGIC EFFECTS

In pharmacologic doses, oxytocin has a variety of effects on many target tissues other than the uterus and the breast. This chapter will discuss the reported pharmacologic effects of oxytocin, but will exclude those effects which are evoked by both physiologic and pharmacologic doses of oxytocin on the uterus and the breast.

The Male Reproductive Tract

Oxytocin (25 - 200 mU/ml) reversibly reduced the noradrenaline-, dopamine- and acetylcholine-induced contractions of the vas deferens and in concentrations of 50 - 200 mU/ml also abolished contractions induced by field stimulation (24). At a higher concentration of 400 mU/ml, oxytocin also abolished the response to potassium. The inhibitory effect of ocytocin was related to the external calcium concentration. Thus oxytocin suppressed the contractile response of the vas deferens, partly by reducing the availability of extracellular calcium.

The reduction in frequency, weight and sperm content of spontaneous ejaculates in male rats after copulation could be partially offset if the animals were injected with 1 U of oxytocin daily for four days, compared to saline-treated animals (3). In particular, oxytocin treatment reduced the interval during which the number of sperm was reduced following two ejaculations during copulation. The mechanism for this action of oxytocin is not known, but its contractile effect on the epididymis and the vas deferens may be important.

Carbohydrates

Vasopressin, angiotensin and oxytocin were able to stimulate gluconeogenesis from lactate by 25 - 50% in hepatocytes from starved rats (398). The minimal effective concentration for oxytocin was 0.2 nM compared to 0.02 pM and 1 nM for vasopressin and angiotensin respectively. Such stimulation of gluconeogenesis was dependent on extracellular Ca^{2+} and was abolished in the absence of the latter. Insulin was able to prevent stimulation of gluconeogenesis by oxytocin but not by vasopressin and angiotensin.

Oxytocin (1 nM - 1 μM) rapidly stimulated glycogen degradation in rat liver hepatocyte suspensions, with the release of glucose, but was 1000-fold less potent than vasopressin (167). The effect of oxytocin on glucose release was critically dependent on extracellular Ca^{2+} since in the absence of the latter in the extra-cellular medium, glycogenolysis continued to be stimulated but glucose was not released into the medium. The phosphorylase A content of hepatocytes was increased by oxytocin (minimum dose of 0.1 nM) and was again dependent on Ca^{2+}. The insulin-like effect of oxytocin appeared in the same range of Ca^{2+} concentrations as the maximum insulin effect on glucose activation (40). Neither calcium ionophore nor verapamil, a calcium antagonist, affected the phenomenon. Hence, it was postulated that calcium ions may act at the membrane transducer level rather than at the hormone receptor level.

Lipids

Physiologic concentrations of vasopressin (20 μU/ml of 80 M) and pharmacological concentrations of oxytocin (100- to 1000-fold higher than vasopressin) were found to stimulate increased incorporation of [^{14}C]acetate into lipids and increased protein accumulation by the WRK-1 cells, a cloned cell line derived from rat mammary tumor (230). Pharmacologic concentrations of insulin (1 μM), which itself was stimulatory to the WRK-1 cells, further enhanced and was synergistic with the effects of the neurohypophyseal hormones. For example, while vasopressin increased incorporation of acetate into lipids by two-fold and insulin by ten-fold, when both were present the increase was 30-fold. There was a latent duration of at least 24 hours before the effect of vasopressin was observed which was unlike other responsive tissues. The lipids synthesized were mainly polar lipids rather than triacylglycerols which were produced by differentiated mammary gland. The effect of vasopressin on the WRK-1 cells appeared to depend on the pressor activity of the hormone since (1-Deaminocysteine, 8-D-arginine)vasopressin, a vasopressin analog that lacked pressor activity, had no effect. Hence this stimulatory effect was only observed in oxytocin given at pharmacologic doses.

Gastric Motility

If further substantiated, oxytocin could be used to restore gastric contractility. Three patients with persistent, prolonged and complete post-vagotomy gastric atony, who did not respond to either conventional treatment with gastric aspiration and intravenous fluids or jejunal feeding, as well as the reported field trials with bethanecol hydrochloride and metoclorpramide, were treated with oxytocin (165). An intravenous dose of 5 - 20 U of oxytocin in 500 ml of normal saline was given every four hours for three days and gastric contractions were screened fluoroscopically. Within three days, effective gastric contractions were seen in one patient, partially effective contractions in another, and only some reverse peristalsis in the third patient who responded effectively after another three days of oxytocin. Side-effects of the treatment included colicky abdominal pain and some bouts of diarrhea, all of which were overcome by reducing the dose of oxytocin. No satisfactory explanation for the efficacy of oxytocin could be advanced except that denervation makes smooth muscle more susceptibe to drugs. This was the basis for using oxytocin in these three patients.

OTHER EFFECTS

Pressor Effects

Phenoxybenzamine, an alpha adrenergic receptor blocker, potentiated the pressor response to neurohypophyseal hormones but not to angiotensin in the adult rat (125). The site and mechanism of this action was elucidated by *in vivo* and *in vitro* studies on isolated aortic strips from the rat. With phenoxybenzamine or phentolamine blockade, the pressor response of oxytocin was enhanced, but with beta blocking agents such as propanolol there was no change in the pressor response of oxytocin. With aortic strips, the potentiating effect of phenoxybenzamine was only seen if norepinephrine was added to the bathing medium. Thus, an interaction between phenoxybenzamine and noradrenaline was probably involved in potentiating the pressor response of neurohypophyseal hormones.

In dogs, it was shown that a normal vascular response to oxytocin did not return even after long-term recovery from sympathetic injury (150). The vascular effect of oxytocin depended not on a normal pattern of peripheral adrenergic innervation but on an unknown, more central activity of the sympathetic nervous system.

The vascular and renal effects attributed to prolactin are probably due to contamination of prolactin preparation by vasopressin and oxytocin (374). Incubating the prolactin preparation with antisera against vasopressin and oxytocin, or with pregnancy plasma which contained enzymes that could destroy vasopressin and oxytocin rapidly, but not prolactin antisera, completely eliminated the antidiuretic, milk ejection and blood pressure activities of prolactin preparations.

Natriuretic and Antidiuretic Effects

Gershkovich (149) postulated that the mechanism of the diuretic actions of the pituitrin, vasopressin and oxytocin was through the activation of the pituitary-adrenal system since these neruohypophseal hormones had no diuretic effect in adrenalectomized or hypophysectomized rats. However, these hormones were still able to increase sodium excretion and some additional mechanism was therefore thought to be involved in their natriuretic effect.

Both antidiuretic and diuretic effects have been ascribed to the action of oxytocin and thus the influence of oxytocin on urine flow in normal and vasopressin-deficient (Brattleboro) rats was investigated (15). In normal rats, oxytocin increased the urine flow, but in Brattleboro rats a consistent antidiuresis occurred. Thus, when endogenous vasopressin was absent, oxytocin showed weak antiduretic action, probably because of a structural similarity to vasopressin. In the presence of vasopressin, oxytocin displayed a diuretic action – probably due to a competitive displacement of the vasopressin, which was a more potent antidiuretic.

The antidiuretic action of oxytocin on the frog urinary bladder was influenced by the mucosal and serosal pH (258). Mucosal acidification to a pH of 6.5 reduced the water permeability by 88%, while alkalinization to a pH of 10.5 increased it by 200%, to the same concentration as oxytocin. Similar effects were seen with serosal pH changes. The effect of the hydrogen ion concentration on the action of oxytocin was probably at a post-cAMP level since it affected the action of cAMP. Freeze-fracture studies showed that an increase in the mucosal pH increased the number of intramembranous aggregates of particles induced by neurohypophyseal hormones.

Although there was a positive qualitative correlation between the hydroosmotic effect of oxytocin and its analogs on the intact bladder of the toad and the activation of cAMP, there was a larger discrepancy between these two activites as the potency of the peptide in generating cAMP increased (188). This was partly because only a small fraction of the cAMP generated was necessary for maximal hormone response.

Mesotocin, an amphibian neurohypophyseal hormone, produced diuresis in low doses in the bullfrog kidney in a dose-dependent manner (256). At higher doses, the diuretic response of mesotocin decreased progressively. The changes in urine flow were directly proportional to the changes in glomerular filtration rate and it was suggested that mesotocin might dilate the afferent glomerular vessels. The diuretic effect of mesotocin was subsequently confirmed in the Chilean toad and the mud puppy with a similar dose-response relationship (142).

Miscellaneous Effects

Oxytocin stimulated the sodium pump of the frog skin epithelial cells. Oxytocin significantly stimulated sodium extrusion either at normal or low potassium in the medium (1). Ouabain or the removal of potassium prevented this action but amiloride was ineffective in this respect.

Oxytocin stimulation of frog skin in a "polarized condition" was attenuated after it had been exposed to lanthanum, a trivalent cation which competed with calcium for its sites in biologic membranes (391). In the "depolarization" site however, the skin could be stimulated for sodium transport by oxytocin even after lanthanum. Under the latter conditions, lanthanum permanently stimulated the short circuit current while in the polarized state it stimulated the short circuit current only transiently. These investigators postulated that oxytocin could control the permeability properties of the apical membrane of the epithelial cells through a general modification of the membrane structure rather than through specific local actions on each specialized permeability mechanism.

When added to the outer solution bathing the frog skin, silver in concentrations of 10^{-4} M increased the electrical conductance and interfered significantly with the action of oxytocin, probably through an interference with several transport processes such as the permeability of sodium entry (205). A large dose of oxytocin (100 ng/g) was required to induce the spawning reflex activity of the teleost killfish (209). This dose was larger than that for arginine vasotocin (47 - 62.5 ng/g), but oxytocin was more potent than the telestean peptide, isotocin, which required a dose of 480 - 790 ng/g, while an equipotent dose of arginine vasopressin was 53 ng/g. It was the vasopressor activity of all of these peptides that induced the spawning reflex.

Oxytocin in doses of 100 mU/day accelerated certain metamorphic events in the larval tiger salamanders *(Ambystoma tigrinum)* (271). Oxytocin could partially block the antimetamorphic effect of prolactin. *In vitro* studies wherein tail tissues were incubated, suggested that the observed effect of oxytocin was diuretic.

Oxytocin increased the beating rates of single myocardial cells and cell clusters of rats and improved the arrhythmia of myocardial cells which showed fibrillation (11). Oxytocin was also found to increase myosin ATPase activity and protein synthesis in myocardial cells cultured with serum-containing media, but not in serum-free media.

■ ■ ■

Chapter 7

■ OXYTOCIN RECEPTORS

In two recent reviews on oxytocin receptors, (333, 336), Soloff summarized the information available up to that time, including many of the findings that have been made in his laboratory. High affinity binding sites for tritiated oxytocin have been detected in particulate fractions from rat uterus, oviduct and lactating mammary gland, myometrium of the sow, ewe and human and ovine endometrium. The binding activity, presumably reflecting oxytocin receptors was localized in the enriched plasma membrane fractions from rat uterus and lactating mammary gland. In isolated enriched cell preparations, the receptors could be demonstrated to be present in myoepithelial cells of the mammary gland. The apparent dissociation constant (K_d) for oxytocin-receptor interaction in different tissue preparations was in the nanomolar range. Apparently the K_d of oxytocin interaction with its uterine receptors was similar to the concentration of oxytocin eliciting half-maximal contraction of rat isolated uterus. Binding was specific for both the target tissue preparation as well as the ligand. Interestingly, the affinity of binding sites for oxytocin analogs corresponded generally to their potencies as agonists or antagonists and some investigators have used such oxytocin receptor-oxytocin analog binding as an additional tool in evaluating the agonist or antagonist properties of the analogs. Divalent metal ions could potentiate oxytocin activity on the receptors and thereby enhanced the sensitivity of the uterus and mammary gland to oxytocin. Uterine oxytocin receptors in some species were regulated by estrogens and progesterone. Administration of estrogens such as diethylstilbestrol increased the sensitivity of the myometrium to oxytocin by increasing the concentration of uterine oxytocin receptors. In contrast, the progesterone-dominated uterus had a low concentration of oxytocin receptors. In the non-pregnant uterus, maximal oxytocin binding to myometrial particulate fractions was seen at estrus, when the uterus was also known to be most sensitive to oxytocin. The extent of endometrial $PGF_{2\alpha}$ synthesis in the ewe closely paralled the increased concentration of oxytocin binding sites. During pregnancy, the rat uterus showed an abrupt and marked rise in its sensitivity to oxytocin which was simultaneously accompanied by an increase in the concentration of uterine oxytocin receptor binding activity. In the rat, there appeared to be some evidence that supported the regulation of oxytocin receptors by estrogens and progesterone as having an important role in the initiation of labor and parturition.

RADIOLABELED OXYTOCIN FOR RECEPTOR STUDIES

A significant difficulty in studying oxytocin receptor or binding sites is the lack of readily available, purified tritiated oxytocin. It appears that iodinated oxytocin loses a significant portion of the ability of the unlabeled oxytocin to bind to its receptor sites. The synthesis of tritiated oxytocin is not easy and consequently most of the investigations on oxytocin receptors have been confined to a few groups. Interestingly, even work from the same laboratory has indicated different sources for the tritiated oxytocin used for various publications.

Iodinated oxytocin can be used as a potential ligand for receptor binding and as an intermediate in the synthesis of tritiated oxytocin (133). Monoiodinated oxytocin ([3-iodo-Tyr^2] oxytocin) had to be freshly prepared prior to bioassay. Diiodinated oxytocin ([3,5-diiodo-Tyr^2] oxytocin) was inactive as an agonist or as an antagonist, but monoiodinated oxytocin was a weak, reversible inhibitor of oxytocin in the uterine assay. Inhibition by monoiodinated oxytocin lacked specificity. Diiodinated oxytocin was catalytically tritiated to produce [^{3}H] oxytocin with a specific activity of 25 - 31 Ci/mmol. Tritiated oxytocin could then be used as the ligand to successfully measure oxytocin receptors in the uterine muscle.

LOCALIZATION AND MECHANISM

The localization of oxytocin receptors was studied using various enzyme markers which were specific to different subcellular fractions of the tissues studied. Crankshaw et al. (83) used 5′-nucleotidase (a plasma membrane enzyme), NADPH-cytochrome C reductase (an endoplasmic reticulum marker enzyme), and succinate-cytochrome C reductase (a mitochondrial marker enzyme), to determine in which subcellular fraction of the rat myometrium the oxytocin receptors were located. The distribution of tritiated oxytocin binding sites paralleled the distribution of 5′-nucleotidase but not the other two enzymes, suggesting that the oxytocin receptor was located on the plasma membrane of the smooth muscle cell of the rat uterus. The binding of tritiated oxytocin to the most enriched plasma membrane fraction indicated specificity when studied with different oxytocin analogs. The binding of oxytocin to plasma membrane fraction of the rat myometrium gave a K_d of 1.98×10^{-9} M and a capacity of 1.28 pmol/mg.

Walter (376) has reviewed the available data concerning identification of the sites of oxytocin involved in the recognition and activation of uterine receptors. Isoleucine in position 3, glutamine in position 4, proline in position 7, and leucine in position 8 are amino acid residues in the corner positions of the β-turns which appear to contain the elements for recognition for the oxytocin receptor. The presence of the aromatic ring of the tyrosine in position 2 appears to enhance the affinity of binding. The hydroxy group in the *ortho* position of the tyrosine side-chain, folded over the 20-membered ring of oxytocin, acting together with the asparagine carboxamide group, appears to be the main active element responsible for initiating the oxytocic response. Conformational considerations rather than structure-activity investigations have been essential to the understanding of the contribution of amino acids, such as asparagine, to the affinity and intrinsic activity of the hormone. For instance, the B—CH_2 group of asparagine, while playing no role in binding, is involved in the stabilization of the peptide backbone structure and thus the ability to bind with the hormone receptor. The available data on whether the peptide backbone contributes functionally to the hormone-receptor interaction are not helpful and therefore, it is unclear whether the C=O and N—H groups of the peptide backbone are involved in the hormone-receptor complex. Thus it appears that available information on the conformational structural aspects of oxytocin indicates that a variety of analogs can be produced by selectively modifying activity profiles with the potencies in one or more systems enhanced or reduced. These could be used to probe the mechanism of oxytocin action.

Several studies have shown that tritiated oxytocin binds nonspecifically to subcellular fractions other than the one containing the bulk of the plasma membranes. Marklee et al. (218) took material from lactating rabbit mammary gland which specifically bound oxytocin and purified it on a sucrose density gradient. This material was determined to be purified plasma membrane fraction. Scratchard analysis of this oxytocin receptor revealed a K_d of 1.83×10^{-9} M and a capacity of 670 fmol/mg protein for the crude fraction, compared to a K_d of 2.8×10^{-9} M and a capacity of 1700 fmol/mg protein for the purified fraction. Treatment of the purified material with different detergents resulted in a loss of binding capacity to $[^3H]$oxytocin but if it was preincubated with $[^3H]$oxytocin prior to detergent treatment, the hormone-receptor complex could be solubilized and remained in the supernant even after centrifugation at 210,000 g for 30 minutes. This solubilized complex was determined with different oxytocin analogs to be oxytocin specific. The detergent-induced change in the binding capacity of the receptor was probably related to a conformational change in the receptor or a displacement of the lipid component of the receptor since it had been suggested that oxytocin receptor has a lipophilic, sterically defined region whose interaction with the side-chain in sequence position 3 made an important contribution to oxytocin binding.

The formation of the oxytocin hormone-receptor complex is believed to lead to membrane depolarisation, an influx of calcium, and increased cytoplasmic levels of Ca^{2+} which activate the contractile proteins. What is poorly known is the precise manner in which the occupied receptors are coupled to the activation of the contractile systems. Using binding studies of oxytocin receptor preparations with tritiated oxytocin, oxytocin receptors in the uterus have been shown to be present in several species including the rat, rabbit, sow, ewe, and women. Nissenson et al. (248) reported their studies demonstrating that in estrogen-dominated rabbits, oxytocin receptors were coupled to uterine muscle contraction. Microsomal membrane from estrogen-dominated rabbit uterus was found to contain high-affinity specific oxytocin binding sites with a K_d of $2 - 3 \times 10^{-9}$ M. Specific oxytocin binding exhibited an optimum pH between 7.5 and 8.0, while Mg^{2+} or Mn^{2+} were necessary for the maximal specific binding, but submillimolar concentrations of Ca^{2+} inhibited specific binding. In progesterone-dominated rabbit uterus, oxytocin binding sites were not detectable in microsomal membranes. Using the above oxytocin receptor binding in estrogen-dominated rabbits, the relative binding and the uterotonic activities of ten synthetic neurohypophyseal hormone analogs were determined. A qualitative correlation was found between binding and uterotonic responses. Angiotensin II and insulin did not compete with uterine oxytocin binding sites. Hence, it was concluded that in estrogen-dominated rabbits, the high affinity oxytocin binding sites had the selectivity for neurohypophyseal hormone analogs expected of physiologic receptors that coupled to uterine contraction. Specific oxytocin binding sites were found in both the microsomal and the mitochondrial fractions, but the concentration of binding sites was lower in the mitochondrial as compared to the microsomal fraction (0.16 vs. 0.37 pmol/mg protein).

Oxytocin binding activity was found to occur in subcellular fractions from the rat lactating mammary gland and in several other species. Such specific tritiated oxytocin-binding activity, reflecting the presence of oxytocin receptors, was localized in the intact myoepithelial cells from the rat lactating mammary gland (302). Enriched myoepithelial cells, obtained by collagenase treatment and sucrose density gradient centrifugation of isolated mammary cells, demonstrated the presence of a single class of specific oxytocin binding sites with an apparent K_d of 5×10^{-9} M. The amount of oxytocin bound at 20°C and at a pH of 7.6 was proportional to the concentration of oxytocin and the number of cells. A steady state was reached by 40 minutes, and 0.45 fmol of oxytocin were bound per 10^6 cells. The K_d of the binding reaction was consistent with a single-step model for the interaction of oxytocin with the binding sites on the cells. Oxytocin analogs competed with oxytocin for the oxytocin receptor sites in the myoepithelial cells in the following order: (deamino)oxytocin > (4-threonine)oxytocin > oxytocin > (0-methylthyrosine)oxytocin > (8-lysine)vasopressin.

(Lysine)-bradykinin and (4-proline)oxytocin did not compete with oxytocin for the receptors. The characteristics of the oxytocin receptors in isolated mammary cells were comparable to those previously described for disrupted mammary cell preparations.

THE REGULATION OF OXYTOCIN RECEPTORS

Nissenson et al. (247) examined the effects of estrogen and progesterone on rabbit myometrial oxytocin receptors in an attempt to explain the effect of these steroids on the sensitivity of the myometrium to oxytocin. The administration of estradiol (200 mg/kg for four days) to rabbits increased the uterine contractile response to both oxytocin and methacholine *in vitro*, but progesterone administration (5.0 mg/kg for four days) after estrogen pretreatment selectively abolished the contractile response to oxytocin but not to methacholine. Measurements of the oxytocin receptors indicated that there was a parallel change in the concentration, but not the affinity, of the oxytocin receptors in the uterine microsomal membranes. Estrogen administration induced an increase in the number of uterine oxytocin receptor binding sites compared to untreated control rabbits, but progesterone reduced the number of binding sites to barely detectable levels. The effect of progesterone in reducing specific oxytocin receptors in estradiol-pretreated rabbits was apparent within 24 hours with short-term (24 hour) treatment with progesterone when specific binding, but not the affinity, was reduced. When actinomycin D, a protein synthesis inhibitor (80 mg/kg, four times) was given for 24 hours, there was a similar reduction in the concentration of receptors as when progesterone (5.0 mg/kg, four times) was given for 24 hours. When both compounds were given together, there was no additive effect. It appeared that estrogens induced an increase in uterine oxytocin receptors in the rabbit, while progesterone, like actinomycin D, acted at the nuclear locus to repress synthesis of oxytocin receptors.

Magnesium, a divalent ion, has been shown to influence the effect of neurohypophyseal hormones and their analogs on the myometrium (264). The effect of some divalent metal ions on oxytocin binding to its receptors in the rat mammary gland membranes was investigated and used to characterize the oxytocin-receptor interaction. Oxytocin binding to its receptors in the rat mammary gland was significantly potentiated in increasing order by zinc, magnesium, nickel, manganese and cobalt, but not by calcium, copper and iron. The affinity constants (K_a) of the oxytocin-receptor binding in the presence of the potentiating divalent ions ranged from 3.1 to 8.6×10^8 M^{-1}. Increasing concentrations of nickel, magnesium and manganese produced an increase in the concentration of binding sites available for oxytocin, but the affinity was not changed. In contrast, increasing concentrations of cobalt increased the affinity of the receptor for oxytocin but not the concentration of available binding sites. Combinations of both types of metal ions were not additive either for the concentration or the affinity of oxytocin-binding sites and therefore probably both types of ions bound to identical receptor sites. The kinetics of the association and dissociation of the oxytocin-receptor complex revealed a fast step for the binding of metal ions followed by a slow, rate-determining step for the binding of oxytocin. Based on these observations, these investigators proposed a model wherein the oxytocin receptor possessed two distinct regions for metal ion interaction. The binding of metal ions to a region designated as region A (availability) produced a receptor with the *maximum number* of available binding sites, but with a *low affinity* for oxytocin. On the other hand, the binding of the metal ion to region B (binding), as with cobalt, resulted in a receptor site of *low availability* but *high affinity*. The binding of divalent metal ions to both sites would result in a receptor of high affinity and full availability. These findings might help to explain some of the effects of divalent metal ions on the action of oxytocin on the uterine muscle *in vitro*, and possibly *in vivo*.

Csaba et al. (86) suggested that neurohypophyseal hormone excess in the neonate resulted in an "amplification" of the hormone receptor. In young rats, vasopressin pretreatment enhanced, while oxytocin pretreatment decreased, the responsiveness of aortic strips to vasopressin. In adult rats, increased sensitivity to vasopressin was seen with both pretreatments.

OXYTOCIN RECEPTORS IN NONPREGNANCY

It is now clear that oxytocin causes the *in vivo* ovine uterus to release prostaglandins but whether this is secondary to mechanically-induced uterine contractions or through an effect on uterine $PGF_{2\alpha}$ synthesis is unclear. Roberts et al. (283) examined the relationship between the estrus cycle, oxytocin-receptor binding and the oxytocin-induced release of $PGF_{2\alpha}$ from the ovine uterus. They found that ovine endometrium incubated *in vitro* released $PGF_{2\alpha}$ spontaneously and oxytocin potentiated this release in a dose-dependent manner. The enhancement of endometrial release of $PGF_{2\alpha}$ with low doses of oxytocin increased with the approach of estrus, reaching a maximum on the day of estrus. Both myometrium and endometrium had high-affinity binding sites (K_d = 5 to 7 $\times$ 10^{-10} M) which reached a peak at estrus, but the binding capacity of endometrium was double that of the myometrium at estrus. Oxytocin potentiated $PGF_{2\alpha}$ release from the endometrium only. Based on their findings, it was hypothesized that the endometrium was a target for oxytocin which interacted with its endometrial receptors to stimulate $PGF_{2\alpha}$ synthesis and that the ovarian steroid hormones might influence this process by regulating the availability of these receptors.

OXYTOCIN RECEPTORS IN PREGNANCY AND PARTURITION

Oxytocin receptors have generated much interest with respect to understanding parturition and uterine sensitivity to oxytocin. Nevertheless, the difficulties in such research have restrained any rapid advances in this area. One such difficulty is the availability of biologically active radiolabeled oxytocin which could be used as the radioactive ligand for the oxytocin receptor. Nevertheless, a few studies have examined the regulation of myometrial oxytocin receptors, the relationship between myometrial oxytocin receptors and intrauterine volume, and between myometrial estrogen and oxytocin receptors during prostaglandin abortion in the rat (7 - 9, 335).

Soloff et al. (335) found that the specific binding of tritiated oxytocin to the myometrial oxytocin receptors of pregnant rats increased signficiantly and rapidly at term, reaching maximum values during labor and parturition. The apparent dissociation constant was 1 - 2 $\times 10^{-9}$ M throughout pregnancy and this was similar to the K_d obtained in estrogen-treated rats. These findings correlated with the marked increase in oxytocin sensitivity of the rat uterus which occurred 6 to 8 hours before term, and which had also been previously noted by other workers. It was suggested that the dramatic and rapid increase in oxytocin binding to myometrial oxytocin receptors may be an important and crucial factor in triggering parturition, besides any increase in circulating oxytocin levels that have been reported by other investigators. The authors postulated that the drop in circulating blood progesterone levels and the steady increase in circulating estrogen levels in the pregnant rat at term, as reported by many, could be responsible for the increase in oxytocin receptor binding activity.

Rat mammary gland oxytocin receptor binding with tritiated oxytocin increased throughout pregnancy, decreased transiently at term and was maximal during lactation (335). This finding correlated well with the sensitivity of the myoepithelial cell to oxytocin as noted by others. However, since the profile of the mammary gland oxytocin receptor binding was not exactly similar to that of the myometrium, the authors concluded that the receptors from the two tissues were differently regulated.

Alexandrova and Soloff (7) showed that the increase in oxytocin receptors began several hours before labor, was maximal at labor, and declined several hours later. In the peripartum and immediately postpartum periods, the change in oxytocin receptors was proportional to the ratio of plasma estradiol to progesterone levels with a correlation coefficient of 0.8 ($p < 0.001$). The increase in oxytocin receptors was preceded, by several hours, by a proportional increase in the concentrations of cytosol and nuclear estrogen receptors. Since estrogens have been shown to increase the concentration of oxytocin receptors in the rat and also in the rabbit, the authors proposed the following sequence in hormonal and receptor changes that occurred with the onset of labor in the rat. The decline in serum progesterone allowed serum estradiol to induce the synthesis of estrogen receptors in the myometrium. The resulting increased concentration of estrogen receptors and their occupancy by estradiol stimulated the appearance of more oxytocin receptors. The increased concentration of oxytocin receptors was then able to trigger coordinated myometrial activity, and thus labor, by interacting with the circulating oxytocin. Certainly this is an attractive hypothesis for the rat, but there is just no published data to support such a sequence in women. Indeed, most studies have been unable to show a decline in serum progesterone at the approach of labor.

What is not clear from the hypothesis is whether the increase in myometrial oxytocin receptors is brought about by the increased ratio of circulating estradiol: progresterone or whether it is secondary to the release of pituitary oxytocin which can be stimulated by estradiol.

The same investigators also examined the effects of intrauterine volume, which is influenced by uterine stretch, on the concentrations of myometrial estrogen and oxytocin receptors in the rat (8). Levels of estrogen and oxytocin receptors were compared in gravid and nongravid uterine horns from unilaterally pregnant rats at or near the time of parturition. There was a greater amount of deoxyribonucleic acid (DNA) and protein in the gravid horn compared to the nongravid horn. The quantity of nuclear estrogen, cytosol estrogen and oxytocin receptors were 7.5, 5.2 and 4.8 times greater, respectively, in the myometria from gravid compared to nongravid horns. However, the concentration of receptors per cell was the same in gravid and nongravid horns at or near parturition. It was concluded that uterine stretch caused hyperplasia and hypertrophy of the myometrial cells with a resultant increase in DNA and protein. However, estrogens are potent stimulants of the myometrium and are able to induce myometrial hyperplasia and hypertrophy. Thus, uterine stretch *per se* may not be responsible for all of the reported observations. Because the marked rise in concentrations of oxytocin and estrogen receptors occuring at the time of labor appeared to be hormonally induced (7), it was concluded that uterine stretch did not have a role in the regulation of myometrial oxytocin receptors.

An attempt was made to determine if hormonal and receptor changes occuring in the rat spontaneous term labor could be made to occur prematurely when pregnancy in that animal was terminated with $PGF_{2\alpha}$ (9). The administration of $PGF_{2\alpha}$ on day 18 of gestation in the rat induced premature delivery on day 20, and was associated with an increase in the concentration of myometrial estrogen and progesterone receptors. Progesterone administration inhibited both premature delivery and the increase in estrogen and oxytocin receptors. In contrast, the changes in receptor and premature delivery did not occur if $PGF_{2\alpha}$ was given on day 15 of pregnancy, when plasma progesterone was near maximal levels. However, if given on day 10, $PGF_{2\alpha}$ increased estrogen and oxytocin receptors by day 11 and caused fetal resorption by day 12. Exogenous progesterone blocked these changes. The effect of $PGF_{2\alpha}$ appeared to be through its luteolytic activity, since exogenous progesterone blocked the $PGF_{2\alpha}$ -induced myometrial receptor changes. Azastene, an inhibitor of progesterone synthesis, produced effects similar to $PGF_{2\alpha}$ on day 10 of pregnancy. The authors were also able to show that the ratio of plasma progesterone to estradiol was inversely proportional to the concentration of myometrial receptors in $PGF_{2\alpha}$ - and azastene-treated rats. Therefore, it was hypothesized that the biochemical events preceding $PGF_{2\alpha}$-induced abortion were similar to those in spontaneous labor in the rat. These include a withdrawal of progesterone followed by an increase in the concentration of myometrial estrogen receptors which in turn, caused the appearance of increased oxytocin receptors leading to the onset of labor when the circulating oxytocin interacted with its receptors.

Most of the investigations on oxytocin receptors have thus far been carrried out in animal species but a report on oxytocin receptors of the human uterus appeared recently. Using a highly specific tritiated oxytocin which was synthesized employing the method of catalytic substitution of halogen for hydrogen, Sakamoto et al. (290) studied oxytocin receptors of the gravid human uterus at different stages of pregnancy and during labor. The tritiated oxytocin had a specific activity of 19 Ci/mM and a biologic activity of 350 U/mg. Maximum uptake or binding of the tritiated oxytocin was in the 20,000 g pellet of homogenized human myometrium. Optimum binding was at a pH of 7.4 at 20°C with an incubation time of 90 minutes, with further augmentation by 10% and 40% in the presence of Mg^{2+} and Mn^{2+}, respectively. Scatchard plot analysis showed that there was a single type of binding site for oxytocin in the human myometrium with a range of total oxytocin concentration from 0.4 to 16 nM. The dissociation constants showed a relative increase as gestation advanced, from 1.66×10^{-8} M in nonpregnancy to 1.64×10^{-8} M in the first trimester of pregnancy and 1.25×10^{-8} M at term. The K_d remained the same with or without labor. In contrast, the number of binding sites showed a relative increase as gestation advanced, from 1.0×10^{-12} mol/mg protein, and 1.1×10^{-12} in the first trimester to 4.7×10^{-12} at term, but declined to 1.6×10^{-12} with the onset of labor. Such a drop was explained by receptor occupancy and saturation with oxytocin when the woman was in labor.

There are independent receptors for oxytocin and catecholamines in the frog skin. Both oxytocin and catecholamines activated adenylyl cyclase and increased cAMP content. Propranolol completely blocked the generation of cAMP induced by catecholamines but had no effect on oxytocin-induced cAMP. Phentolamine enhanced cAMP production by catecholamines, and oxytocin combined with isoproterenol had an additive effect in increasing cAMP (300).

OXYTOCIN ANALOGS AND RECEPTORS

Competitive inhibition of oxytocin binding to its receptors has been employed to determine the potency of oxytocin analogs as antagonists of oxytocin-induced uterine contractions. Four antagonists, [N-acetyl, 2-0-methyltyrosine] oxytocin; [1-(β-mercapto-β, β-diethyl propionic acid)] oxytocin; [1-(β-mercapto-β, β-pentamethylenepropionic acid)] oxytocin; and [1-(deaminopenicillamine), 4-threonine] oxytocin were evaluated for their ability to compete with tritiated oxytocin for binding sites in particulate fractions from rat uterine homogenates (334). The apparent dissociation constants of these antagonists were not significantly different from the K_d values as calculated from their individual potency as an antagonist to oxytocin-induced uterine contractions. Thus, the binding sites for oxytocin were part of the receptor complex and with oxytocin, "spare receptors" did not appear to be present in significant amounts. The relative potency of each antagonist appeared to be dependant on its affinity for the receptor rather than its intrinsic activity.

Oxytocin analogs substituted at the p- and m-positions of 2-tyrosine and/or at the N-terminal amino group were known to be competitive inhibitors of neurohypophyseal hormones. The pA_2 values, which were a measure of the binding of a number of substances to the oxytocin receptor, were analyzed for the 1,2-substituted oxytocin analogs (272). The pA_2 values of these analogs suggested that there was a significant resonance effect of p-substituted groups in the 2-tyrosine position when the hormone bound to its uterine receptors, but the N-terminal amino group exerted less clearly characterized effects.

Deaminooxytocin had an extremely protracted action on the *in vivo* rat uterus preparation with a response-length increment of 26 ± 4 minutes, compared to 3.5 ± 0.4 minutes for similar doses of oxytocin (146). Since the plasma half-lives of these two peptides were similar, it was concluded that the protracted uterine action, deaminooxytocin, was due to specificially bound peptides which remained in the uterus long after unbound peptides had been cleared from the circulation. However, the affinity of deaminooxytocin for the uterine receptor was not markedly different from that of oxytocin.

■ ■ ■

Chapter 8

▪ OXYTOCIN-NEUROPHYSIN INTERACTION AND OXYTOCIN METABOLISM

OXYTOCIN-NEUROPHYSIN INTERACTIONS

It is now generally accepted that oxytocin and vasopressin are transported in the axons from the hypothalamus to the posterior pituitary, bound to their respective specific carrier proteins, the neurophysins. Oxytocin is bound to neurophysin I while vasopressin is bound to neurophysin II. Both neurohypophyseal hormones are thought to be released into the circulation together with their neurophysins, and some investigators have even employed this as their premise for conclusions about oxytocin or vasopressin release by measuring the corresponding neurophysin. Apparently there are differences in the binding of oxytocin and its neurophysin compared to vasopressin and its neurophysin. A number of factors governing oxytocin neurophysin interactions have been elucidated and the molecular mechanism of such interactions have been proposed.

Cohen et al. (76) have reviewed in detail the biochemical aspects of neurophysin-neurohypophyseal hormone complexes and challenged the earlier concept that neurophysin is a "carrier protein" or an inert byproduct of the biosynthesis of neuro-hormones. There is general agreement that under standard conditions of pH 5.7, one oxytocin molecule is bound per neurophysin chain of 10,000 daltons and that there is some discrete degree of positive cooperativity at low values of saturation as reflected by binding isotherms. The binding of oxytocin to neurophysin appears to depend significantly upon neurophysin dimerization and accounts for some of the discrepancies in the reported values of the apparent K_a from different laboratories. There are two classes of binding sites on the neurophysins, with cooperative interaction between them. It has been suggested that the binding system can now be envisioned as an allosteric dimeric neurophysin molecule, with four pre-existing, structurally related sites. Apparently, saturation of two of the four binding sites weakens binding to the other two sites.

Changes in Neurophysin

Bovine neurophysins tend to associate in solutions to form dimers. Nicolas et al. (244) have proposed two possible models for the binding of oxytocin to its neurophysin and there has been experimental support for both models. In the first model, the monomer is in equilibrium with the dimer from, with one site per monomer and one site on each protomer of the dimer available for oxytocin, including the possibility of cooperative interactions between the two dimeric sites. In the second model, which is supported by better fit for the entire binding isotherm at various neurophysin concentrations, the equilibrium equation took into consideration the possible pre-existent, or ligand-induced, isomerization of the dimer. The second model is able to tell us that there is a positive cooperative interaction between the two sites on the dimer such that binding to the first site ($K_2 = 1.3 \times 10^5\ M^{-1}$) on the dimer quadruple the affinity for the second ligand molecule bound ($K'_2 = 5.3 \times 10^5\ M^{-1}$).

Smythies et al. (330) analyzed the detailed molecular mechanism for oxytocin-neurophysin I binding. They suggested that oxytocin fitted snugly into a deep cavity in the neurophysin, so that the complex formed a globular, water-extruding mass. The mode of interaction was determined by binding various amino acid terminals of one molecule to those of another. The differences between neurophysin I and neurophysin II were apparently related to the amino acid differences between oxytocin (ile at 3) and vasopressin (phe at 3).

Oxytocin and analogs bearing the oxytocin residue in position 3 preferentially enhanced the formation of dimer complexes within neurophysin I. Binding studies strongly suggested that Tyr-49 was not directly involved in the first, higher affinity oxytocin binding, but was associated in some fashion with the second weaker site (244). Further analysis with neutral solvent pertubation, ultraviolet spectroscopy showed that Tyr-49 of the neurophysin was either 15% protected from solvent in the native neurophysin I or that Tyr-49 transitions were only very slightly shifted compared with the monomer forms. Thus, these investigators proposed that the unmasking of Tyr-49 consecutive to binding with oxytocin was not due to neurophysin dimer formation but, rather, that Tyr-49 was reoriented upon oxytocin binding to the first site, making the second, weaker site available.

The Tyr-49 of neurophysin and the tyrosine of oxytocin were studied by fluorescence during neurophysin-oxytocin interaction (310). Binding increased the fluorescence intensity of Tyr-49 by 13%, while the fluorescence of tyrosine in oxytocin was almost completely quenched. Less than half of the fluorescence of Tyr-49 was lost by transfer to the quenched hormone tyrosine. The distance between Tyr-49 and the tyrosine of hormones bound to the strong hormone binding site was greater than 5 Å. Further enzymatic and chemical cleavage of neurophysins indicated that the conformational information necessary for the transition resided within the amino acid sequence adjacent to Tyr-49. When oxytocin was bound to neurophysin, the tyrosine ring of the oxytocin underwent hindered rotation, the rotation at 20°C being 130 s^{-1} and at 42°C, 900 s^{-1} (31).

Measurements by temperature-jump relaxation of the neurophysin-oxytocin interaction showed general agreement with circular dichroism and pH titration studies which indicated that the interaction occurred between a negatively charged carboxyl group on the neurophysin and a protonated α-amino group on the hormone (263). The findings also showed that oxytocin bound with greater affinity to neurophysin in the dimer form than the monomer form. Further studies of the interaction of bovine neurophysin with oxytocin using nuclear magnetic resonance confirmed that there was a strong interaction of the amino group with the negatively charged moiety on the neurophysin and that this occurred in the hydrophobic region which was rather inacessible to solvent (33).

Studies examining the optical activity and peptide affinity of bovine neurophysin and oxytocin indicated that the peptide-binding of neurophysin diminished between pH 6 and 2 in a manner suggestive of salt-bridge formation by the peptide (44). Nevertheless, there was significant binding with a constant binding affinity between pH 2 and 1 which might represent binding in the absence of salt-bridge formation. However, deamino-oxytocin did not effectively compete with oxytocin at low pH, indicating that significant conformational differences existed between these two peptides or that binding at low pH was still dependent on salt-bridge formation. Because of this, Breslow and Gargiulo (44) considered the possibility that the decrease in the binding affinity between pH 6 and 2 was the result of conformational change rather than salt-bridge formation.

^{13}C-enriched oxytocin derivative, [9-(2-^{13}C)-glycinamide] oxytocin (derivative 1) and [1-hemi-L-(2-^{13}C)-cystine] oxytocin (derivative 2), interacted strongly with neurophysins I and II but [1-hemi-D-(2-^{13}C)cystine] oxytocin did not (29). The binding of neurophysins with derivative 1 had more motion than with derivative 2, but binding with the latter produced a greater chemical-shift difference. The glycinamide-9α carbon derivative 1 was in *fast exchange* in the hormone-neurophysin complex while the half-cystine-1α carbon of derivative 2 was in *slow exchange.* Using similarly ^{13}C-enriched oxytocin and oxytocin derivatives, Blumenstein et al. (30) went on to show that the interactions of the side-chain and disulfide moieties of the hormone with the neurophysins did not change as a function of pH. At neutral pH, there was a slow exchange with the neurophysins but at low pH there was a fast exchange with a dissociation rate constant that was 100-fold greater than at neutral pH, and was probably due to breaking of the salt-bridge involving the N-terminal amino group of oxytocin and a side-chain carboxyl group of neurophysin. At high pH, the bound hormone was always in slow exchange, even when the binding affinity was decreased as a result of deprotonation of the α-amino group of the free hormone.

[1-Hemi-D-(3-^{13}C)cystine] oxytocin was found not to interact with neurophysin *in vivo* (100). Increasing the concentration of neurophysin in the presence of this analog led to aggregation of the system.

Changes in Oxytocin

The interaction of oxytocin with bovine neurophysin I (Np I) and neurophysin II (Np II) was studied using nuclear magnetic resonance spectroscopy and observing the tyrosine-2 aromatic protons of the hormone (32). This method was used because of conflicting earlier studies that suggested that at neutral pH, the bound hormone was in slow exchange, while others found that there was a fast exchange. The present investigators found that at neutral pH, most of the observations suggested that the bound hormone was in slow exchange. Using different isomers, they found that [2-L-(2-^{13}C)tyrosine] oxytocin and [2-L(2-^{13}C)tyrosine,8-arginine] vasopressin interacted strongly with both Np I and Np II with a 1:1 of hormone to protein stoichiometry. [2-D(2-^{13}C)tyrosine] oxytocin did not react with either neurophysin. Based on the small rate constant and the substantial chemical shifts associated with the binding process, it was concluded that a substantial conformational change occurred at the tyrosine-2 position when the hormones bound to neurophysins.

Using the same carbon-13 nuclear magnetic resonance spectroscopy [^{13}C-Leu^{3}] oxytocin, [^{13}C-Gly^{9}] oxytocin and [^{13}C-Ile^{3},Gly^{9}] oxytocin, it was demonstrated that the residue 3 (and not 9) was altered upon binding to neurophysin, suggesting that residue 3 was involved in the oxytocin-neurophysin interaction (156).

Using selectively enriched ^{13}C-labeled proline-7 and ^{13}C-labeled leucine-8 oxytocins as probes, oxytocin interaction with bovine neurophysin I was analyzed with nuclear magnetic resonance spectroscopy (79). The findings indicated that the Pro^{7} peptide bond remained in the trans conformation upon complex formation with the neurophysin, and the segmental mobility of the entire tripeptide tail of the oxytocin was unaffected upon binding to the neurophysin. No large conformational rearrangement occurred in the region of the tripeptide tail of the oxytocin.

OXYTOCIN METABOLISM

Pharmacokinetics

Using the curve-fit method for estimating plasma half-lives from vasopressor responses in the rat, the half-lives of vasopressin and oxytocin analogs which were substituted in 1, 1 and 6, and 8 positions were determined (145). Deamino-dicarba analogs of arginine vasopressin and arginine vasotocin were found to be long-acting to the same degree, respectively, as deamino-vasopressin and deamino-vasotocin. Deamino-vasopressin and 1-hydroxyoxytocin were not longer-acting than vasopressin or oxytocin. Three 8-substituted analogs of oxytocin, (8-ornithine)-oxytocin, (8-glutamine)-oxytocin and (8-proline)-oxytocin were long-acting. Thus, deamination of arginine vasopressin and arginine vasotocin prolonged their half-lives significantly, but this was not so for the deamination of oxytocin. Hydroxylation of the 1 position for arginine vasopressin, but not for oxytocin, prolonged its half-life. Substitution in the 8-position of oxytocin with arginine, ornithine, glutamine or proline also prolonged the half-lives.

The metabolic clearance rates of oxytocin in the pregnant ewe and fetus, in the pregnant baboon, and in women were reported (91, 97, 151). In the chronically catheterized pregnant sheep, measurements of oxytocin by RIA showed that baseline fetal plasma oxytocin levels were significantly higher than in the maternal plasma (1.6 ± 0.13 versus 1.1 ± 0.13 μU/ml). Using continuous infusion of oxytocin to achieve a steady state, the mean fetal metabolic clearance rates of oxytocin were 12.0 ± 1.35 and 12.1 ± 1.09 ml/kg/min at oxytocin infusion of 64 and 640 μU/kg/min. The mean metabolic clearance rates of oxytocin in the pregnant ewe (125 - 140 days gestation) were 12.1 ± 2.6 and 12.4 ± 1.4 ml/kg/min, respectively, with infusions of 80 μU/kg/min. Thus, the pregnant ewe and the fetus had similar metabolic clearance rates of oxytocin near term. Because the authors were unable to obtain any uterine contractions with fetal oxytocin infusion, and only with very high doses of maternal oxytocin infusions, they concluded that oxytocin did not appear to cross the placenta in either direction. It is noteworthy that one of the investigators had previously shown that oxytocin given to the fetal sheep was able to raise maternal intrauterine pressures and uterine contractions, and that oxytocin could, therefore, cross the placenta. In the pregnant woman, there is data to indicate that oxytocin can cross the placenta from the mother to the fetus at term (95).

In the pregnant baboon at term (171 - 179 days), the apparent half-life of oxytocin was 1.1 and 1.7 minutes in two animals, with a late half-life of 9.9 and 17.3 minutes, respectively (91). One animal had an apparent half-life of oxytocin of 9.9 minutes. The clearance rates were 3.1, 3.2 and 11.7 ml/kg/min in the three baboons, with calculated production rates of 97, 74 and 390 pg/kg/min, respectively [2 pg = 1 μU oxytocin]. Thus, the pregnant baboon had a lower metabolic clearance rate than the pregnant ewe at term. It appeared that the pregnant baboon had two components to the curve of disappearance of oxytocin, an initial rapid component and a slow secondary component. When a bolus dose of oxytocin was injected into the fetal cord vessels and oxytocin levels were measured serially from different maternal sites, such as the femoral vein and amniotic fluid, the findings suggested that oxytocin could probably cross the placenta from the fetus to the mother.

In man, pharmacokinetic studies of oxytocin employing steady state infusions and using RIA for oxytocin determinations in the plasma indicated that the half-life of oxytocin was influenced by the dose of oxytocin infused (97). With a steady state infusion rate of 256 mU/min of oxytocin, the half-life of oxytocin was 8.9 ± 1.6 min, but decreased to 16.7 ± 0.8 min with an infusion rate of 132 mU/min. This finding was similar to that previously reported in the rat. The metabolic clearance rate of oxytocin in women was 21.5 ± 3.3 ml/kg/min, which paralleled the combined blood flow volumes of the liver and the kidneys, thus supporting the role of these two organs as the major sites for clearing oxytocin. The apparent volume of distribution of oxytocin was 302 - 308 ml/kg, indicating that oxytocin was distributed into a space or spaces other than the circulating plasma volume.

Catabolic Enzymes

The properties of the enzyme(s) involved in the metabolic degradation of oxytocin have not been fully elucidated. Two enzymes acting on the linear portion of the oxytocin molecule have been identified in the cytoplasmic fraction of chicken liver (50). They are carboxyamidopeptidase (releasing $Gly.NH_2$) and prolyl peptidase (releasing $Leu\text{-}Gly.NH_2$). Using CM-Sephadex C-50 and DEAE-Sephadex A-50 chromatographic separation, some workers were able to purify carboxamidopeptidase 134-fold with a yield of 23%, and prolyl peptidase 71-fold with a yield of 20% (50). The specific activity of the final preparations were 181 μU/mg protein for carboxamidopeptidase, and 96 μU/mg protein for prolyl peptidase. The optimum pH for carboxamidopeptidase was 6.0 - 6.5, and for prolyl peptidase activity it was 7.5. Carboxyamidopeptidase activity was inhibited by Mn^{2+}, Zn^{2+}, Ca^{2+}, Co^{2+}, and stimulated by EDTA. Prolyl peptidase was inhibited by Zn^{2+} and Mn^{2+}. The K_m value of both enzymes for oxytocin was 1.5 - 2.4 μM. It is noteworthy that while the kidney is the principle site of oxytocin degradation in mammals, the liver is the major site of neurohypophyseal hormone inactivation in the chicken. With oxytocin analogs, which were protected against reduction and aminopeptidase cleavage, the enzymic cleavage of the linear chain of the peptide molecule by carboxamidopeptidase and prolyl peptidase, was probably the only possible way that these analogs were inactivated.

More recently, Simmons and Walter (316) reported on the purification and characterization of carboxyamidopeptidase, an enzyme which inactivated oxytocin. The enzyme was purified 3800-fold with a 22% yield from toad skin, a target tissue for oxytocin. The purified enzyme was able to inactivate arginine vasopressin and oxytocin by hydrolyzing the Arg^8-Gly^9-NH_2 and Leu^8-Gly^9-NH_2 bonds respectively. The enzyme had both trypsin and chymotrypsin-like activities. It was maximally active at pH 7.5 to 8.5 for vasopressin, and pH 7.0 for oxytocin. Carboxamidopeptidase could be inhibited by a variety of agents, including trasylol, concanavalin A and sulfhydryl group reagent. Inhibition of carboxyamidopeptidase by p-chloromercuribenzoate could be reversed by cysteine. The molecular weight of the enzyme was about 100,000 and it was made up of two identical subunits of 48,000 daltons. Each subunit consisted of a heavy chain of 28,000 daltons and a light chain of 19,000 daltons joined by a disulfide bond. The active site of the enzyme was in the heavy chain.

Human colostrum was shown to have low levels of oxytocinase activity, but was rich in nonspecific aminopeptidase activity since it degraded oxytocin, arginine vasopressin and lysine vasopressin, and cleaved L-leucyl-β-naphtylamide (386). The nonspecific aminopeptidase levels were not reduced by preincubation of the colostrum with anti-oxytocinase. It was suggested, based on previous reports, that human colostrum degraded oxytocin – probably via an initial reduction of the disulfide bond. Inactivation of the neurohypophyseal hormones was also found in rhesus monkeys. The enzymes in the colostrum were thought to be from enzymes present in the mammary gland.

Metabolic Degradation

Analogs of oxytocin and deaminooxytocin with 4-glutamine replaced by 4-glutamic acid methylester readily lost their uterotonic activity when incubated with rat serum. This was presumed to be due to hydrolysis of these compounds to the less active 4-glutamic acid derivatives. Pliska and Rudinger (274) studied the inactivation of oxytocin, desaminoxytocin (DOT) and their 4-glutamic acid methyl ester in the rat serum and uterus. In the rat serum, the analog deamino-[Glu(OMe)^4] oxytocin (DGM-OT) was rapidly converted to the slightly active demethylated product, deamino-[Glu^4] oxytocin, but the uterus inactivated DGM-OT more slowly than it did oxytocin. Hydrolysis of the ester bond did not seem to influence the overall rate of inactivation, although DGM-OT was disposed more rapidly from the uterus than DOT. The ratio of uterotonic *in vivo/in vitro* activities for both [Glu(OMe)^4] oxytocin (GM-OT) and DGM-OT was low (0.07) compared to the high ratio of 2.7 for DOT (assuming a ratio of 1.0 for OT). The responses to GM-OT and DGM-OT were clearly shorter than those to OT, rather than it being the case of the effect of DOT being noticeably prolonged. Such results from these *rapidly inactivated* analogs could perhaps provide support for some predictions of the three-compartment model for tissue distribution of neurohypophyseal hormones. The three compartments are: 1) an initial compartment comprising of blood plasma and a part of the extracellular space; 2) the receptor compartment and; 3) the residual compartment which comprises the remaining tissues in which the hormones are distributed.

The renal tubular handling of oxytocin *in vitro* appeared to be different from that of angiotensin II (268). Microperfusion of radiolabeled angiotensin II and oxytocin through rabbit kidney tubule segments revealed that hydrolysis of oxytocin was not observed in either the proximal straight or cortical collecting tubule segment, and that the rate of reabsorption was low, in contrast to that of angiotensin II which was rapidly hydrolyzed and reabsorbed.

■ ■ ■

Chapter 9

▪ THE OXYTOCIN CHALLENGE TEST

Since the suggestion by Hammacher in 1966 that the response of the fetal heart rate to oxytocin-induced contractions in late pregnancy would be a valuable method for predicting the ability of the fetus to withstand the stress of labor, the oxytocin challenge test (OCT) has been extensively evaluated. The principle of this test is to produce a relative hypoxic stress to the fetus by inducing uterine contractions and then to observe the fetal response. The end point observed is the fetal heart rate and pattern. Thus, it is essentially a placental respiratory reserve test or placental oxygen transmission reserve test. If this feature of placental function is compromised, the fetus will be unable to cope with the relative hypoxia of the uterine contraction, and changes in fetal heart rate or pattern will result. Thus, a positive test has been defined as a uniform deceleration of the fetal heart rate that reflects the wave form of the uterine contraction, with the onset at or beyond the peak of a contraction and occurring repetitively. A negative test, therefore, would be one where no change in fetal heart rate was observed with uterine contractions.

The OCT is performed with the patient in the semi-Fowler position with a slight left lateral tilt to minimize the risk of supine hypotension (which can reduce uterine blood flow and therefore increase the false positive rate). Blood pressure is monitored at frequent and regular intervals. Uterine contractions should be monitored with an external tocodynamometer and the fetal heart is monitored using either the Doppler ultrasound, phonocardiography or abdominal fetal electrocardiography. At the start of the test, a 10-minute baseline tracing is obtained. Baseline fetal heart rate variability and periodic fetal heart rate decelerations and accelerations are looked for. If there are no uterine contractions or less than three uterine contractions per ten-minute period, then oxytocin is administered intravenously by a constant infusion pump starting at 0.5 mU/min and the dose gradually increased at regular intervals (every 20 minutes) until regular uterine contractions at the rate of three contractions every ten minutes are obtained. When an adequate contraction pattern is established, oxytocin is discontinued and monitoring continued until uterine activity returns to baseline.

Several reviews of the OCT pointing out its indications, results, reliability and problems were published in the last few years (112, 179), 311, 389). At the same time, numerous publications on the OCT also began to appear (19, 20, 25, 87, 111, 143, 173, 189, 203, 251, 304, 338). Most of these were based on clinical experience with varying numbers of the OCT, but several aspects of the test became increasingly focused on. These included classification of the abnormal fetal heart rate and pattern obtained, what constituted an abnormal test, the high false positive results found by some, the fetal outcome, and finally, complications arising from the test.

OCT and Fetal Well-Being

OCT was found to be of good prognostic value in several reports of the experience at different centers. In one experience with 1,827 OCTs on 1,366 patients from the 34th week of pregnancy onwards, using 5 mU oxytocin/min, perinatal mortality was 13.2/1000 (111). Normal or negative OCTs were seen in 85.4%, prepathological response in 7.8%, and pathological response in 6.8% of the patients.

In another report on 537 OCTs done on 364 high-risk pregnancies, there was a good correlation between a positive OCT and poor perinatal outcome (20). Although this study confirmed the reassurance of a negative OCT for a further week of intrauterine survival in 98.2% of cases, one-third of their positive OCT results were false positive when further tested by spontaneous labor.

The ominous implication of a positive OCT was further confirmed where late decelerations during labor, meconium staining of amniotic fluid and cesarean section rates were highest, compared to those with a negative or a suspicious OCT (25). However, the experience of the authors indicated that vaginal delivery can be undertaken without endangering the fetus if fetal scalp blood pH measurements can be performed.

The OCT was used for antepartum fetal testing together with biochemical surveillance in 278 high-risk patients who had 420 OCTs (19). Those with a positive OCT had their pregnancies terminated. Perinatal outcome in patients with a positive OCT was significantly worse than in patients who had a negative OCT. Suspicious OCTs frequently became positive OCTs and such patients were more likely to bear growth-retarded infants. A negative OCT generally had a favourable outcome. There was inconsistent correlation of estriol excretion with the OCT. Four prenatal deaths occurred, but all were considered to be unpreventable.

That a negative OCT will not result in a fetal death within a week was further confirmed in 1,209 OCTs carried out on 533 fetus or 529 pregnancies at risk for uteroplacental insufficiency (173). The incidence of negative OCTs was 72%. In 69 fetus with a positive OCT, 47 were subjected to the stresses of labor with maternal hyperoxygenation and lateral positioning. Twenty of the 47 tolerated this without evidence of fetal distress. During a positive test, reactivity of the fetal heart rate significantly increased the likelihood that a fetus would tolerate labor. The authors' experience with this group of patients showed that fetal and perinatal mortality rates in the patients identified as at risk for uteroplacental insufficiency and assessed with the OCT were no greater than in a comparable group of pregnancies without identifiable risk for uteroplacental insufficiency.

The OCT was compared to urinary estriol in the management of high-risk pregnancies in 362 patients (87). Perinatal mortality was six times higher with a positive OCT than with a negative one. Perinatal mortality was also six times higher if expectant treatment rather than immediate delivery was undertaken with a positive OCT. A negative OCT had a 0.3% risk of death within a week of the test. Low estriols combined with a positive OCT gave a poor prognosis but normal estriol levels were associated with good perinatal outcome if expectant treatment was given for a positive OCT.

In a restrospective analysis of patients (430) who had OCTs, compared to stillbirths that did not have an OCT done, it was concluded that the OCT was preventing stillbirth and avoiding unnecessary premature intervention (143). There was only one stillbirth in the OCT group, whereas 35% of the patients with stillbirths had indications for an antepartum OCT but did not have it. The OCT might have prevented those deaths.

In an attempt to quantify the OCT it was found, from OCTs performed on 350 women, that if contractions were produced at the rate of three per ten minutes, the intrauterine pressure profile mimicked normal labor and the first fetal heart rate deceleration was likely to occur between the second and third effective contractions (304). A fetal heart rate deceleration/contraction ratio of more than 10% could significantly predict an abnormal fetal heart rate during parturition and an increased chance of cesarean section.

Spellacy et al. (338) compared serum human placental lactogen (hPL), serum-free estriol and OCTs in 149 high-risk women who were more than 34 weeks pregnant. A positive OCT was found in 15.4%, and serum hPL had a significant correlation with serum estriol. Significantly lower hPL levels were found in the positive OCT groups. Patients within the low hPL subgroup had a significantly higher incidence of a positive OCT (42.9%) than the high hPL subgroup (80%). In contrast, serum estriol was unhelpful in predicting the OCT result since estriol levels were similar in the positive and the negative OCT groups, and the low estriol subgroup had the same frequency of positive OCTs as did the overall group. Hence, it was recommended that a serum hPL done at 34 weeks gestation could screen pregnant women for fetal welfare and if hPL level was low (4.0 μg/ml), an OCT was indicated, since a positive OCT yield would be higher in this group.

The experience with 163 OCTs performed at the Maryland General Hospital confirmed the conclusions that have already been made in larger series (203).

When the OCT was compared to amniocentesis for the management of patients with prolonged pregnancy, it was found that the incidence of meconium was similar and that the presence of meconium was of little value in the management of prolonged pregnancy (189). Instead, the OCT, being simpler, less invasive and carrying fewer complications than amniocentesis, could effectively monitor fetal well-being. Apparently three times as many inductions of labor were performed based on meconium findings in the amniocentesis group compared to the positive OCT group.

OCT with Buccal Oxytocin

A variation of the OCT was to employ buccal pitocin instead of intravenous oxytocin. Sixty-three buccal pitocin challenge tests were carried out on 40 women with high-risk pregnancies and then compared to the intravenous oxytocin challenge test (181). The starting dose of buccal pitocin was 100 U and doubled every 20 minutes until 800 U or three uterine contractions per ten minutes were established. Interpretation of the changes in the fetal heart rate and pattern were similar to those for the intravenous route. The average duration of the test with the buccal pitocin was 120 minutes, and uterine activity returned to normal within 20 minutes of removal of the buccal pitocin. There was one false positive out of 63 buccal pitocin challenge tests, and the buccal route was satisfactory in 31 of 40 patients in establishing uterine contractions. The authors claimed that the advantage lay in the administration of buccal pitocin. The use of buccal pitocin, in spite of achieving satisfactory physiologic labor-like concentrations of oxytocin, is currently not favored as the absorption rate via this route is not constant and therefore carried potential problems.

Boehm et al. (36) compared buccal and intravenous oxytocin and oral PGE_2 for use in the contraction stress test. The results showed that in a dose of 0.5 mg that was increased to 1 mg one hour later, PGE_2 was generally ineffective in initiating contractions and hypertonicity. Also, side-effects render PGE_2

unacceptable for the test. Buccal oxytocin was more effective but the incidence of hypertonicity (3.3%), the doses of buccal tablets required, and the higher success rate with intravenous oxytocin made buccal pitocin unacceptable.

OCT Versus The Nonstress Test

Both the OCT and the nonstress test (NST) have been used for antenatal fetal evaluation and then compared. In one study, 1,000 NSTs and 919 OCTs, performed on 362 high-risk pregnancies were compared, and the last tests (performed within a week of delivery) were compared (278). Of the 300 NSTs analyzed, 11 were nonreactive with 55% fetal morbidity, 18 were equivocal with 44% fetal morbidity, and 271 were reactive with 28% fetal morbidity. Of the 278 OCTs studied, 17 were positive with 65% fetal morbidity, 42 were equivocal with 31% fetal morbidity, and 219 were negative with 27% fetal morbidity. A combined nonreactive NST and a positive OCT were associated with a fetal morbidity of 83%. The experience of these authors indicated that an OCT was slightly better than the NST as identifying fetal morbidity, but that if the equivocal NST were included, then NST was better than the OCT. In another, smaller pilot study comparing NSTs and OCTs in 83 patients who were going to have induction of labor, it was found that both tests were predictive of fetal performance in labor (219). A somewhat high false positive rate of 20% was found with OCTs but the series was too small.

When the NST was compared to the OCT, it was found that if fetal heart acceleration occurred following fetal movement, the OCT was negative in 99.5% of cases, while a positive OCT occurred in 12% if the NST showed abnormal fetal activity (190). Thus, a normal NST obviated the need for an OCT which could be more selectively employed.

Fetal Heart Rate Reactivity and the OCT

To improve the predictive value of the OCT for fetal outcome, Lin et al. (205) proposed a combination of the OCT and fetal heart rate reactivity classification. The results of the OCT were then classified into four groups; positive nonreactive, positive reactive, negative nonreactive, and negative reactive. Of the 293 patients having OCTs, there were six perinatal deaths – all occurring in the two groups with a nonreactive pattern. A nonreactive pattern was defined as diminished or decreased events of fetal heart rate acceleration to less than three occasions per 30 minutes. The positive nonreactive test had 90% accuracy in predicting a compromised infant while a negative reactive test assured good fetal outcome.

An analysis of 27 patients with positive OCTs and their outcome of monitored labor gave similar findings about the additional predictive value of fetal reactivity when combined with the OCT (39). Those with a reactive pattern during the OCT (15 patients) had a good fetal outcome, although 8 of the 15 showed distress during labor. A combination of a positive OCT and a nonreactive baseline fetal heart rate uniformly ended in fetal distress during labor.

Similar observations about the additional value of fetal heart rate acceleration were made on 1,570 OCTs where a positive OCT was more associated with a lack of fetal heart rate acceleration and gave more compromised fetuses than if there was good fetal heart rate acceleration (127). Another recommendation was that loss of beat-to-beat variability and a negative OCT should be interpreted as an ominous prognostic sign since the loss of beat-to-beat variability suggested a complete inability of the fetus to react to any stimulus (171).

Four cases of a rare, atypical heart pattern found during the oxytocin challenge test at 31 - 35 weeks gestation in hypertensive gravidas were reported (21). The heart rate pattern showed a loss of baseline variability and repeated variable decelerations, some with a late deceleration component. All of the patients had oligohydramnios and fetal growth retardation; three of the fetuses died.

There have been many definitions of the NST. In an attempt to define the criteria for the NST which were the most predictive and therefore reduced the need for an OCT, a definition of the NST based on previous criteria described by different authors was proposed (10). It was proposed that the most apparently predictive criteria for the NST should be based on a fetal heart rate acceleration of five in any 20-minute period with an increase of 15 beats/min lasting 15 seconds or more with the test conducted for 40 minutes. If such criteria were employed, the OCT was required after the NST in only 14% of cases.

False Negative OCTs

It has been previously indicated that a negative OCT ensured a fetal survival for seven days, but reports of false negative OCTs served to emphasize the need for cautious interpretation of the test.

Many reports of false OCTs continued to be published and the clinical circumstances under which a false negative OCT may occur continue to expand (106, 120, 141, 209, 217, 222, 238, 254, 257, 291). Together, these reports involved 13 pregnancies and unfortunately, all the fetuses died (291). The types of gestational complications in which a false negative OCT were reported included diabetes mellitus, hypertension, prolonged pregnancy, anemia with sickle cell disease, and even an abdominal pregnancy (254). It was clinically more reasonable to have diagnosed the abdominal pregnancy by some of the obvious physical signs and an OCT would not have been necessary (194). In one patient with Class A diabetes mellitus where fetal death occurred four days after the last negative OCT (41 weeks), it might be questioned whether it was not unusual to allow a diabetic pregnancy to go beyond 40 weeks gestation, especially when her plasma estriol was falling from 10 ng/ml to 5.5 ng/ml at 41 weeks (141). Neuhoff et al. (238), in reviewing the literature up to 1978, found 6,593 negative OCTs of 7,903 OCTs carried out on 4,589 patients. The incidence of false negative OCTs was 0.349% (23 cases). False negative OCTs could be categorized into three major groups: 1) fetal demise that cannot be predicted by an OCT since fetal death resulted from factors other than deterioration of uteroplacental respiratory reserve (such as intrapartum insults); 2) misinterpreted OCT results such as an obviously positive OCT presented as a negative one (291); and 3) a truly false negative OCT in which fetal death was the result of a rapid, acute deterioration of uteroplacental respiratory function.

Clearly, false negative OCT results are going to be encountered with fetal death occuring in those pathophysiologies which are unrelated to a gradual chronic compromise of placental respiratory reserve. However, similar false negative NST results have also been reported (389).

PITFALLS

Interpretations of fetal heart rate pattern during the OCT are open to subjectivity by the analyst, although the biophysical evaluation of the fetus is supposed to be based on objective criteria. In a blinded study, it was shown that there was considerable disagreement between five physicians, subspecializing in maternal-fetal medicine, in their individual evaluations of 50 OCTs (265). On the average, any two physicians would agree only 52% of the time. There was also considerable disagreement on their evaluation of fetal heart rate reactivity patterns. When the majority of the physicians did agree on the OCT reading, the Apgar scores and the incidence of growth-retarded infants reflected the seriousness of the tracing. Thus, it appeared that in the distinctly positive OCT, there was no problem with the interpretation, but in those with minimal or subtle changes, disagreement was greater. Nevertheless, it is in such a group of patients that a discriminating test is needed.

Another difficulty was the incorrect interpretations of fetal heart tracings in OCTs, as was pointed out (390). Such erroneous interpretations of fetal heart tracings were noted to be present even in publications (291).

One minor but practical problem was the variability in paper speeds, and therefore the accompanying errors in comparing OCT tracings of different speeds. This was well brought out by several criticisms (105, 390) drawing attention to a publication which had nonstandardized paper speeds (291).

ADVERSE AND LONG-TERM EFFECTS

Hyperstimulation of the uterus occurred in 2.8% of 2,792 OCTs (252). Seven patients went into labor within 24 hours of hyperstimulation. Two infants, each weighing less than 2,500 gm, one small-for-dates and the other a trisomy-18 syndrome, were delivered. No harmful effects were noted from the hyperstimulation.

Dominguez et al. (11) classified fetal heart rate abnormality pattern of unintentional hyperstimulation during the oxytocin challenge test as "prepathological." Such abnormalities constituted 8% of all their OCTs, and included late and variable decelerations, baseline tachycardia, transient bradycardia and loss of beat-to-beat variability. A significantly higher incidence of intrapartum meconium and fetal distress was found in the transient bradycardia prepathological group.

A case of inadvertent induction of antepartum hemorrhage in a woman with unsuspected placenta previa was described and attention was drawn to this as a potentially serious complication of the OCT (241).

The long-term effects of repeated intrauterine exposure to oxytocin for the OCT were reported from retrospective and prospective studies. In a study which repeatedly compared 54 women who had had OCTs to 60 women who did not, there was no significant difference in the Brazelton Neonatal Behavioral Scale in the offspring during the first 12 hours or three days of life (294). Jaundice and respiratory distress were not any more frequent after exposure to the OCT. Thus, the OCT itself did not appear to have any demonstrable adverse effects on an otherwise healthy fetus.

The evaluation by clinical, biochemical and behavioral methods of infants up to 1 year of age, whose mothers had had the OCT, showed that infants with a positive OCT had poor state organization and reflexive performance when compared to those with a negative OCT (295). Those with a positive

OCT also showed evidence of intrauterine malnutrition, but not greater asphyxiation than those with a negative OCT. Such findings were not altogether surprising since a positive OCT implied placental respiratory insufficiency which could affect fetal neurobehavioral and growth development.

■ ■ ■

Chapter 10

▪ INDUCTION OF ABORTION AND LABOR

The use of oxytocin for the induction and augmentation of labor and for the reduction of postpartum hemorrhage is well-established in obstetrical practice. Oxytocin has also been used to induce abortion in midtrimester pregnancy but large doses are required. The introduction of specific abortifacients initially offered the possibility of eliminating the use of oxytocin for midtrimester abortion. However, it is clear from the literature that the trend towards emptying the uterus more rapidly in midtrimester abortion has led to the use of megadose oxytocin in combination with one or more abortifacients. Drugs that affect myometrial contractility in pregnancy, including oxytocin, and their use in obstetrics have been recently reviewed for the clinician (353). The side-effects, either from inappropriate or appropriate use of oxytocin, continue to draw our attention to known adverse effects.

INDUCTION OF ABORTION

In an attempt to shorten the induction-abortion interval, many clinical trials were undertaken to determine if the addition of intravenous oxytocin to the intra-amniotic administration of abortifacients would indeed accelerate the abortion. In 295 midtrimester abortions with intra-amniotic urea and simultaneous intravenous infusion of oxytocin at 15 mU/min, the mean induction-abortion interval was 22.3 hours, which was stated to be reduced although no control values were mentioned (328). There was an abortion failure rate of 12.8%, and while water intoxication was not seen, free water clearance and urinary output were decreased during the oxytocin infusion. Apparently such a reduction was not seen if the oxytocin infusion rate was 83 mU/min until 15 hours after infusion. Sandstrom et al. (292) were unable to find any difference in the mean induction-abortion interval in midtrimester abortions performed with intra-amniotic hypertonic saline even when oxytocin (60 U) was infused simultaneously with the intra-amniotic injection. However, the rate at which the oxytocin was given and the total dose delivered were not clearly stated. This is obviously important since the midtrimester uterus is relatively insensitive to oxytocin and an adequate megadose is necessary. Using an intravenous oxytocin infusion of 40 mU/min starting immediately after intra-amniotic instillation of 20 mg of $PGF_{2\alpha}$ and varying volumes and concentration of hypertonic saline, Adachi et al. (2) were able to induce 309 midtrimester abortions within a mean time of 15.5 to 22.2 hours. The shortest time taken was when 50 ml of 20% sodium chloride was used. The dose of oxytocin used did not produce any evident side-effects. However, side-effects from the hypertonic saline and the $PGF_{2\alpha}$ were seen.

The extra-amniotic instillation of Rivanol (6,9-diamino-2-oxyethyl acridine lactate) followed by intravenous oxytocin (70 U over 12 hours) after 24 hours was found to be effective in midtrimester abortion (253). $PGF_{2\alpha}$ given by itself in an initial dose of 0.25 mg followed by 0.75 mg every two hours, or half of this, or in combination with Rivanol did not produce any significant advantage over Rivanol alone.

The induction of abortion using oxytocin combined with other abortifacients for cases of missed abortion, missed labor and intrauterine fetal death were also reported (12, 28). In one report, eight patients with missed labor, seven patients with missed abortion longer than four weeks, and one patient with intrauterine fetal death at 40 weeks had uterine activity induced by the extra-amniotic instillation of 1000 - 1500 ml of normal saline at 8 - 18 drops/min (28). If contractions were not established, intravenous oxytocin was infused at a rate of 7 mU/min. All patients aborted in 3 - 17 hours, with a mean of 9.1 hours. Only eight of the women needed 5 - 10 U of oxytocin. In the other report, the abortion was induced with: 1) extra-amniotic instillation of 200 μg PGE_2 followed by 100 - 200 μg hourly and intravenous oxytocin infusion starting at 32 mU/min and doubled every half hour, or; 2) extra-amniotic instillation of 500 μg of PGE_2 initially and repeated at 4, 6 and 8 hours, and intravenous oxytocin started after 6 hours (12). All of the 12 patients having midtrimester abortions aborted with a mean induction-abortion time of 24.8 hours with the first regimen. The mean induction-abortion interval with the second regimen was 11.0 hours. The second case involved six patients — three midtrimester abortions, two missed abortions and one hydatidiform mole.

Since calcium is a crucial ion involved in the mechanism of muscle contraction, one investigation employed an intra-amniotic instillation of 16-phenoxy-ω-tetranor-PGE_2-methyl sulfonamide (sulprostone) (SHB 286) alone or in combination with intravenous oxytocin or intra-amniotic calcium gluconate (362). In the 90 patients studied, the combination of 3 mg sulprostone with

2.75 mg calcium gluconate gave a higher abortion rate (87%) within 30 hours, followed by the combination of 2 mg sulprostone and intravenous oxytocin with a 78% abortion rate within 30 hours. The dose of oxytocin used was not stated.

Laminaria Tents

There has been much interest in the use of laminaria tents as an adjunct to other uterotonic agents in the induction of midtrimester abortions. The rationale behind this adjunctive therapy with laminaria tents is to soften the cervix thus reducing cervical resistance to strong uterine contractions. Thus, the induction-abortion interval and risk of cervical injuries or complications secondary to cervical dystocia may be reduced. The use of laminaria tents for the induction of abortion is not new and was mentioned in many standard gynecology textbooks in the 1960s. Many studies were carried out using laminaria tents before, with, or after the administration of either prostaglandin, urea, hypertonic saline or intravenous oxytocin for the induction of midtrimester abortions (160, 172, 279, 345, 402). Strauss et al. (345) studied 80 patients at 14 - 20 weeks' gestation undergoing a midtrimester termination of pregnancy. The termination was initiated with 20 mg of prostaglandin $F_{2\alpha}$ and 80 mg of hyperosmolar urea given intra-amniotically and followed by an intravenous infusion of oxytocin at 333 mU/min. One group of patients received no further therapy, a second group had laminaria tents inserted into the cervix at the time of the abortifacient injection and left there for four hours, and a third group had the laminaria tent placed four hours before the injection and removed at the time of the abortifacient injection. The insertion of the laminaria tent four hours prior to injection was significantly effective, reducing the mean injection-abortion interval to 12.9 hours as compared to 17.7 and 20.8 hours in the other groups. There was no demonstrable increase in complications with the laminaria tent than without.

Horowitz (172) studied 100 midtrimester abortions using 40 mg of intra-amniotic $PGF_{2\alpha}$ followed by the insertion of one or more laminaria tents and the intravenous infusion of oxytocin at 166 mU/min. He had no failures and the mean injection-abortion interval was 15.7 hours. Six patients, however, required a reinjection of 20 mg of $PGF_{2\alpha}$ at 24 hours, 19 required removal of the placenta after four hours, and seven had the placenta actively removed in less than four hours. The regime had an advantage in that the laminaria did not have to be inserted before the intra-amniotic injection and a lower dose of oxytocin was needed, although the injection-abortion interval was reduced compared to the study by Strauss et al. (345). However, a higher dose of $PGF_{2\alpha}$ was used.

In contrast to these two studies, the induction-abortion interval was reduced even further, to a mean of 12.5 hours, in 638 hypertonic saline instillation abortions with pre-insertion of the laminaria into the cervix 2 - 3 hours earlier, followed by intravenous oxytocin infusion of 417 - 555 mU/min after the saline instillation (160). The complication rates apparently compared favourably to those reported in a national study by the Center for Disease Control of traditional saline abortion procedure. Thus, it appeared that the use of laminaria to soften the cervix and the megadose oxytocin infusion to stimulate uterine contractions were useful adjuncts in the management of intra-amniotic saline and prostaglandin abortion.

Wilson (402) shortened the induction-abortion interval for midtrimester pregnancies even further and suggested that the regime could be used as a method of terminating midtrimester pregnancy in an ambulatory setting. Abortions were performed on 177 midtrimester pregnancies using either 40 mg of $PGF_{2\alpha}$, 40 mg of urea, or a combination of 40 mg urea and 20 mg $PGF_{2\alpha}$ given intra-amniotically. All of the patients had laminaria tents inserted into the cervix before the intra-amniotic injection. One hour after intra-amniotic instillation, 50 U oxytocin in 100 ml intravenous fluid was given as an intravenous infusion, but neither the rate of infusion nor the total dose of oxytocin used was given. The injection-to-abortion interval was reduced to a mean of 8 hours 42 minutes when urea and prostaglandin were combined, compared to 12 hours 32 minutes with prostaglandin alone, and 21 hours 14 minutes with urea alone. Apparently the reduced dose of $PGF_{2\alpha}$ (20 mg) also reduced the incidence of side-effects due to $PGF_{2\alpha}$. Thus the synergistic effect of urea and prostaglandins was further enhanced by laminaria and oxytocin.

Complications

While pharmacologic attacks at multiple target sites undoubtedly enhanced the speed of inducing midtrimester abortion and thus became more widely employed, it was to be expected that major complications would arise. Two cases of uterine rupture were reported following midtrimester abortion by laminaria, $PGF_{2\alpha}$ and megadose oxytocin (279). Interestingly, both patients were older (34 and 37 years), multiparous women. In one case the rupture was on the lower posterior wall of the uterus, and in the other it was through the left lateral part of the lower uterine segment into the broad ligament. In both patients, the abortion time took longer than the reported average time taken. Because the ruptures were above the cervix, the authors felt that the use of oxytocin was responsible and compared their cases to those of uterine rupture in the third trimester. Nevertheless, they felt that the safest course for midtrimester abortion in such women continued to be with $PGF_{2\alpha}$ and laminaria but with a reduced dose of oxytocin infusion. The doses of oxytocin used in these two patients were, however, 24 mU/min and 120 - 240 mU/min, considerably lower than those used in other reports.

Complications of abortions induced by oxytocin-augmented intra-amniotic instillation of abortifacients were reported (157, 267). There were two cases of uterine rupture associated with oxytocin-augmented saline-induced midtrimester abortion, both ruptures occurring anteriorly, starting from close to the cervix upwards, one on the left and one on the right (157). Unfortunately, one patient died. The authors emphasized that the possibility of uterine rupture should be considered whenever oxytocin is used to shorten abortion times. Perry et al. (267) reported that 5 out of 80 (6.2%) nulliparous women suffered cervical lacerations as a result of midtrimester abortions induced with intra-amniotic $PGF_{2\alpha}$ and megadose intravenous oxytocin. One laceration extended to the lower uterine segment of the corpus and another was associated with myometrial necrosis caused by cornual sacculation and ischemia. No multiparous women sustained uterine injury. Thus, midtrimester abortion induced by intra-amniotic $PGF_{2\alpha}$ was associated with a significant risk of uterine trauma in the nullipara, and this risk appeared to be increased with the use of oxytocin. With intravenous oxytocin for the termination of missed abortion and missed labor, there was a slight activation of the fibrinolytic and coagulation systems and increased platelet aggregation (45).

INDUCTION OF LABOR

Several papers have appeared on the combined use of prostaglandins and oxytocin for the induction of labor. Such a combination stemmed from the original report that it produced synergistic effects on uterine contractions. Baxi et al. (22) reported a double-blind study in 50 patients, using a low-dose intravenous oxytocin infusion (starting at 2 mU/min and doubling every hour) or $PGF_{2\alpha}$ (2.5 μg/min and doubling every hour) for the induction of labor for elective or high-risk indications. Both agents were effective and neither generated any major complications in the mother or infant. The maximum oxytocin infusion rate required was 16 mU/min, with a maximum mean infusion rate of 9.6 mU/min. In another double-blind study comparing $PGF_{2\alpha}$ and oxytocin for the induction of labor in 28 gravidas at term (399), the findings indicated that both agents were equally effective. The dose of oxytocin used was 1 mU/

min for half an hour, increasing to 2 mU/min for half an hour, 4 mU/min for 1 to 2 hours, and 8 mU/min for 7 to 8 hours. A clinical trial of oral PGE_2 versus intravenous oxytocin for the induction of labor in 98 patients showed that there was no significant difference in the duration of labor with either agent but that the success rate was 82% with PGE_2 compared to 65% with oxytocin (237). However, there were remarkably more patients with unfavourable cervix and with Bishop scores of less than 5 given oxytocin than those receiving PGE_2. This could have influenced the outcome of induction. Hypertonus was seen in one case each with PGE_2 and oxytocin. Nausea and diarrhea were more frequent with PGE_2. The dose of oxytocin used was 2 mU/min initially, and doubled every hour up to a maximum of 16 mU/min or establishment of adequate uterine contractions.

Kierse et al. (187) compared oral PGE_2 and intravenous oxytocin for the induction of labor in hypertensive pregnancies. Oxytocin was initially given at 2 mU/min and the dose was doubled every 20 minutes until adequate contractions were obtained. They found that in both nulliparae and multiparae the mean duration of labor was shorter with intravenous oxytocin than with oral PGE_2 but that the need for analgesia was apparently greater with oxytocin induction than with PGE_2. With oxytocin, the duration of labor was shorter by 25% in both multiparae and nulliparae.

A similar observation was made by Lykkesfeldt and Osler (1979) on the efficacy of intravenous oxytocin compared to oral PGE_2 in inducing labor. In 325 patients, labor was induced by using 0.5 mg PGE_2 orally administered every half hour (maximum 500 U/day), or primary amniotomy and automatic oxytocin intravenous infusion with a Cardiff pump. The results showed that 100% of the patients given intravenous oxytocin who had a Bishop score of 0 - 6 delivered within 24 hours, compared to 45% and 74% of the patients given oral PGE_2 who had Bishop scores of 0 - 6 and 7 - 12, respectively, and 41% given desamino oxytocin (Bishop scores of 0 - 6) who delivered within 48 hours. However, there was a significantly higher incidence of alterations in fetal heart rate which necessitated assisted delivery in patients receiving intravenous oxytocin.

Two schools of thought exist with regard to the dose of oxytocin infusion for inducing uterine contractions; the physiologic dose and the pharmacologic dose. Many clinicians prefer to start with a low dose of oxytocin which is more physiologic. Toaff et al. (356) induced labor in 134 patients with pharmacologic doses of oxytocin (2.6 mU/min and increasing step-wise to 422.4 mU/min) after amniotomy and compared it to a similar group of 144 patients who were given only physiologic doses (2.6 to 13.2 mU/min). They found that pharmacologic doses of oxytocin gave better results with reduced induction-delivery intervals, incidence of failed inductions and puerperal morbidity. The incidence of hypertonus was similar in both groups irrespective of the oxytocin dose and it appeared that recognizing the "damping sign" in the intrauterine pressure curve preceeding uterine hypertonus might help to prevent overt uterine tetany. Certainly in the United States an important change in the labeling information for oxytocin mandated by the FDA included starting with a low dose and then increasing the oxytocin infusion (see below).

A novel approach to the administration of oxytocin for the induction of labor is to give it as a "pulsed" infusion. Most pituitary hormones are secreted episodically or in "spurts" or pulses. Based on the frequency of detection, Gibbens and Chard had concluded that oxytocin was released in spurts and this has subsequently been established in a minute to minute measurement of oxytocin during pregnancy and labor (96). Therefore, Pavlou et al. (262) carried out a randomized, controlled trial in which the induction of labor at term was successfully accomplished in 12 patients by amniotomy and intermittent oxytocin infusion (pulsed). This was compared to 16 patients who received continuous oxytocin infusion. Both groups had similar induction-delivery and induction-to-full-dilation intervals, but the pulsed group received a significantly lower total dose of oxytocin. Pulsing was achieved using the Cardiff Infusion System Mark III with oxytocin infused for one minute out of every ten. Certainly this is an attractive and physiologic way of administering oxytocin, with a reduced total dose and perhaps a reduction in the adverse effects of oxytocin which are often dose-related or due to inappropriate use.

In a placebo-controlled trial, Golbus and Creasy (152) found that oral PGE_2 was effective in improving the Bishop score and in inducing labor prior to induction with intravenous oxytocin. However, the procedure did not increase the incidence of successful induction in the 50 pregnant women at term that were studied.

A long-acting carba-analog of oxytocin was tried for the induction of labor in 13 pregnant cows near term (371). Four cows received 20 mg dexamethasone 48 hours prior to receiving an intravenous injection of 5 mg of a long-acting oxytocin analog, 1-desamino-1monocarba-[2-Tyr(OMe)] oxytocin (dCOMOT), into the jugular vein. The other nine cows received no dexamethasone and a second dose of dCOMOT was given six hours after the first one if delivery had not occurred. Labor was initiated in all cases with the first dose of dCOMOT and intermittent spurting of colostrum was noted. With dexamethasone pretreatment, the mean duration of labor was 4.35 hours, compared to dCOMOT alone with a mean duration of labor of 14.25 hours. No uterine tetany or tachyphylaxis was observed and the calves were healthy.

Uterine Activity in Induced Labor

The accurate assessment of uterine activity is important in order to avoid hyperstimulation of the uterus. Several studies have examined the quantitative and qualitative aspects of uterine activity in labor induced with oxytocin or prostaglandins and compared them to those of spontaneous labor (138, 207, 208, 288, 308, 404). Woolfson et al. (404) described the development of a simple electronic system that could quantify uterine activity during labor. Using an open-ended, saline-filled polythene catheter inserted into the uterine cavity and connected to the electronic system, the uterine activity was quantified by summing the area under the contraction curve. With oxytocin infusion of 2 mU/min and increasing by 2 mU/min every 15 minutes until there was evidence of hyperstimulation, they were able to confirm that oxytocin-induced labor showed uterine activity that could be divided into an incremental phase after which there was a stable phase with no further change in uterine activity. The incremental phase lasted 1½ to 2 hours; the stable phase 3½ to 4 hours. In spite of a continuing increase in oxytocin infusion, there was no overall increase in uterine activity during the stable phase of labor. Their earlier observations based on a restrospective analysis were now confirmed employing a more sophisticated "on line" electronic system. They felt that this system would enable the selection of the lowest dose of oxytocin with which to produce the optimal uterine activity in each patient. These investigators are currently further evaluating an automatic oxytocin infusion system based on this concept.

It is reassuring that cardiotocographic monitoring of the uterine contraction pattern and their effect on the fetal heart rate in women undergoing $PGF_{2}\alpha$-induced (3 - 15 μg/min intravenously) and oxytocin-induced (2 - 16 mU/min intravenously) labor did not differ from those of spontaneous labor (207). During labor the upward slope of the contraction wave increased steeply with increasing intensity of contraction. Although a high frequency of typical contraction patterns were found in 10 of the 78 patients studied, these could not be attributed to either the oxytocin or the $PGF_{2}\alpha$. Incoordinate uterine contractions were associated with a longer duration of labor and a tendency to more pronounced acidosis in the infant at birth. The authors extended their observation by examining the characteristics of the uterine activity and their effect on the progress of labor (208). They found that the latent phase was shortened by 50% in induced labor (both $PGF_{2}\alpha$ and oxytocin) compared to spontaneous labor but that the active phase of labor was unaffected. The intensity and frequency of the contraction as determined by the Montevideo Units were similar in both spontaneous and induced labors. The variability of intervals between contractions and amplitudes of contractions were likewise similar in spontaneous and induced labor with no overall tendency towards greater regularity seen during the progress of labor. The progress of labor was slower and more uterine work was required if irregular contractions were found when active cervical dilatation had begun. Roux et al. (288) used $PGF_{2\alpha}$, PGE_2 and oxytocin to induce labor in 101 patients for medical reasons. They found that all three agents were efficient in inducing labor and that the uterine contractility patterns and baseline intrauterine pressure were similar with all the three drugs.

In contrast to these reports, the findings of other investigators suggested that oxytocin-induced contractions were not necessarily similar to those of spontaneous labor. The differences in the specific characteristics of the uterine pressure wave-forms of spontaneous labor, oxytocin-corrected hypercontractility and oxytocin-induced hypercontractility clearly indicated that uterine contractions in spontaneous labor and in oxytocin-augmented labor were not identical (307). In oxytocin-treated patients the uterine contractions showed disproportionately higher rates of rise of pressure regardless of the presence or absence of polysystole or of abnormal elevation of the resting pressure. The start-up time, that is, the time required to reach the maximum rate of pressure, was reduced in oxytocin-induced hypercontractility. The authors, however, pointed out that in the absence of abnormal resting pressure or polysystole, there was no evidence that the increased rate of rise of pressure was a threat to the fetus. This study was subsequently expanded to analyze the intrauterine pressure wave-forms of active first stage labor from patients with spontaneous labor by comparing them to those of $PGF_{2\alpha}$- or oxytocin-induced labor (308). While many of the mathematically analyzed characteristics of the uterine contraction pressure wave-forms were similar in spontaneous and $PGF_{2\alpha}$-induced labor, they were different in oxytocin-induced labor. They found that oxytocin-induced uterine contractions had an elevated initial pressure (19.6 ± 8.4 torr), compared to spontaneous labor (11.5 ± 6.0 torr), had shortened duration of contractions, and a reduced ratio of maximum pressure amplitude to maximum rates of pressure rise. The authors concluded that oxytocin-induced uterine contractions appeared to have an altered inotropic state, different from that in spontaneous or $PGF_{2\alpha}$-induced labor. It should be noted that the dose of oxytocin used was not given and that the oxytocin-treated patients included oxytocin-induced labor, oxytocin-corrected hypocontractile labor, and oxytocin-induced hypercontractile labor. It was therefore difficult to ascertain if such characteristics applied equally to oxytocin augmentation with very low or physiologic doses as opposed to pharmacologic doses.

In a retrospective analysis of 2,834 monitored gravid women with singleton pregnancies producing mature infants with vertex presentation, it was reported that patients given $PGF_{2\alpha}$ or intravenous infusions of oxytocin for the induction or augmentation of labor had significantly more tetanic contractions, baseline hypertonus and coupling of contractions (136). Variable and late fetal heart rate decelerations were seen significantly more often in women receiving oxytocin. The authors pointed out some of the shortcomings of this report of their experience, including the fact that patients requiring uterotonics were different from those going into spontaneous labor. Apparently many of the patients undergoing induction were elective (80%) and poor documentation suggested that the uterotonic agents were given, in many cases, merely to augment a labor progressing normally. The doses of oxytocin used were not analyzed. These considerations are important if a determination is to be made whether the properly indicated use of oxytocin produces uterine contractions significantly differently from those of spontaneous labor.

Oxytocin-Induced Labor and Hormone Levels

The effect of induced labor on maternal hormonal changes was reported in two studies (51, 169). Prelabor relaxin levels were similar in women with $PGF_{2\alpha}$ and oxytocin induction of labor. Neither method of induction was associated with any significant elevation of maternal serum relaxin levels (169). Thus, unlike the pig where circulating relaxin levels were increased during labor, this study was unable to define the role of relaxin in induced human labor. During labor induced by amniotomy and intravenous oxytocin, the ratio of maternal serum progesterone to estradiol-17B fell throughout labor, and this was advanced as a basis for uterine activation and sensitivity of the uterus to oxytocin (51). Serum human placental lactogen and pregnancy specific B_1 glycoprotein showed no specific change from late pregnancy. Estradiol-17B rose slightly in early labor, followed by a fall in the second stage of labor, while unconjugated estriol and 11-hydroxycorticosteroids increased progressively and progesterone decreased progressively throughout labor. In spontaneous labor the changes in progesterone to estradiol ratio around the onset of both term and preterm labor have not been conclusively proven and the findings were controversial. Certainly, if the observations that estrogen stimulated an increase in the concentration of oxytocin receptors and that progesterone suppressed them in rats and rabbits (Chapter 8), are applicable to women, then this might explain the change in sensitivity of the uterus at the approach of term.

A single intramuscular injection of 40 or 60 U of oxytocin at about 340 days gestation (one month before the expected date of delivery) could induce delivery of a normal foal by the mare (168). The changes in estrogen, progesterone and total corticoid levels during and after oxytocin-induced parturition were similar to those reported for normal spontaneous delivery in the mare.

Water Intoxication

Several cases of water intoxication arising from high-dose oxytocin infusion continued to be reported (119, 223, 233, 402). The oxytocin-induced water intoxication was believed to be due to a combination of the antidiuretic effect of oxytocin (which becomes apparent with large doses), and the large volume of water sometimes used to administer the oxytocin. In one report, four patients developed hyponatremia and water intoxication during the induction of abortion (233). The common features in all these patients were the large doses of oxytocin (342 - 1282 U) and the infusion of large volumes of saline-free or saline-deficient fluids such as 5% dextrose in water. Two of these patients went into coma with serum sodium levels of 110 mmol/L after receiving 800 U of oxytocin with 7500 ml of 5% dextrose, and 488 U of oxytocin with 6600 ml of 5% dextrose in a short period of time accompanied by low urine output. The second patient was comatose for several days, had hemiparesis which resolved, but had residual homonymous hemianopia and mild dysphasia. The other two patients had hyponatremia with serum sodium of 130 and 126 mmol/L, respectively, but without symptoms. The consistent feature appeared to be megadoses of oxytocin combined with the infusion of large volumes of sodium-deficient fluids. The safety of megadoses of oxytocin for midtrimester abortion was questioned by some readers lest others accept such doses as safe (135). One patient, who had received 1282 U of oxytocin, but only 1900 ml of intravenous fluids, did not go into a coma and had less hyponatremia than the two comatose patients (233). Therefore, low urine output in such a patient, especially in the presence of a large volume of infused fluid, should alert the physician to the possible development of water intoxication.

In another report (233), three patients in spontaneous labor at term received low doses of oxytocin (less than 10 mU/min) to augment uterine contractions, but developed hyponatremic fits. It should therefore be remembered that the risk of water intoxication is not totally removed with low doses of oxytocin either.

In three other reports, each involving a case of water intoxication, all of the patients developed epileptiform convulsions (119, 182, 403). In all the reports, the patients had induction of abortion, one with 295 U of oxytocin and 4650 ml of 5% dextrose (119), another with intra-amniotic $PGF_{2\alpha}$, intravenous oxytocin (141.6 U) and 7 liters of 5% dextrose (182), and the third with extra-amniotic $PGF_{2\alpha}$ and 240 U of oxytocin given in 6850 ml of 5% dextrose and water (403). Serum sodium was 188 mmol/L with a serum osmolality of 247 mOsmol/kg in one case (119), a serum sodium of 126 mmol/L six hours after the fit was reported in the second case (182), and 110 mmol/L shortly before the convulsion in the third case. While the first case took a few days to recover fully (119), another case still had speech and motor control impairments after 10 weeks (182).

These sporadic case reports of water intoxication do give rise to concern. They serve as a reminder that if megadoses of oxytocin are necessary, this complication should be considered. The risk can be minimized by giving intravenous fluids that contain saline, keeping the total volume of fluid used to a normal 24-hour intake, and checking serum electrolytes and osmolality – especially when the urine output does not keep pace with fluid input. These reports, together with those of transplacental hyponatremia (Chapter 9), serve to draw attention to the more frequent occurrence of asymptomatic hyponatremia with the administration of oxytocin than has thus far been recognized.

Other Adverse Effects

Another adverse effect of the induction of labor was the report of fetal circulatory collapse during induction with amniotomy and intravenous oxytocin infusion in a patient with epilepsy (387). However, it is worthy to note that the patient also had anemia. This adverse effect may therefore not be solely due to the undesirable effect of oxytocin as the inducing agent *per se*, although the authors did feel that fetal-maternal transfusion had previously been demonstrated with oxytocin-induced labor. Nevertheless, it cautioned the careful consideration of the need to induce labor with oxytocin in an epileptic woman who is receiving anticonvulsant agents. Other more obvious and clinically well-known adverse effects of induction of labor, such as failed induction and premature induction due to incorrect dates, continue to be emphasized in the literature (80).

Pulmonary edema which occurred during emergence from anesthesia for the termination of a pregnancy has been described in a patient (277). In addition to the reflex adrenergic stimulation, the antidiuretic effect of oxytocin was thought to have contributed towards the development of the pulmonary edema. The patient had received 7.5 U oxytocin.

Given as a dilute intravenous infusion of even 80 mU, oxytocin did not significantly alter the hemodynamics of the systemic and pulmonary circulations in anesthetized, healthy adult females (306). However, if given as a bolus, 10 U of oxytocin induced a drop in femoral arterial pressure of 40%, systemic resistance of 59%, and pulmonary resistance of 44% within 30 seconds of injection in these women. Thus, it is safer to give oxytocin in diluted form intravenously to women who are under anesthesia.

Restrictions on the Use of Oxytocin

The Federal Drug Administration (FDA) issued new labeling restrictions on injectable oxytocin based on the recommendations of its Fertility and Maternal, Health Drugs Advisory Committee (239). Oxytocin should no longer be used for *elective* induction of labor, but remains an important drug for *inducing labor for medical reasons.* Because of some growing concerns about the risk of oxytocin to the mother and fetus in the elective induction of labor, the benefit of which is purely convenience, the FDA held advisory and open meetings on the use of oxytocin. Based on these hearings, labeling for all oxytocin products will now include the following: "(Name of drug) is indicated for the medical rather than the elective induction of labor. Available data and information are inadequate to define the benefits-to-risks considerations in the use of this drug product for elective induction. Elective induction of labor is defined as the initiation of labor for convenince in an individual with a term pregnancy who is free of medical indications."

The known adverse effects of oxytocin-induced labor include uterine hypertonicity, uterine rupture, and fetal and neonatal bradycardia, while the risks of elective induction include inadvertent preterm delivery with an increased risk of respiratory distress, central nervous system damage and death. However, it was equally emphasized that the benefits of oxytocin-induced labor for certain medical problems threatening the health or life of the mother and/or the fetus, such as in rhesus isoimmunization, maternal diabetes, preeclampsia, and premature rupture of the membranes, far outweigh any of the adverse effects of oxytocin. Important changes in the labeling include: 1) intravenous route as the only acceptable route of administration for stimulation of the uterus; and 2) that the initial dose should not be more than 1 - 2 mU/min with gradual increments of no more than 1 - 2 mU/min until a uterine contraction pattern familiar to normal labor is established. All patients receiving intravenous oxytocin must be under continuous observation by trained personnel, and a physician qualified to manage any complications should be immediately available. Some of the recommendations are not new and constitute the requirements imposed in many good teaching medical centers, but the abuse of oxytocin by some physicians probably necessitates that such restrictions be put in writing by a regulatory agency. Since the FDA bulletin in 1978, a new FDA bulletin in June 1981 has announced that oxytocin citrate buccal tablets have been evaluated to be safe and effective for the induction of labor and treatment of hypotonic uterine contractions.

THE THIRD STAGE OF LABOR

It is generally agreed that blood loss and postpartum hemorrhage are reduced with oxytocics given for the third stage of labor. Thus, most of the studies have focussed on reduced side-effects from the oxytocics that have been employed. Sorbe (337) reported that in a two-year period, both ergometrine (0.2 mg) and oxytocin (10 U) given intravenously had comparable hemostatic efficiency but that oxytocin was better in promoting placental separation and expulsion with less risk of placental retention and trapping. Moir and Amoa (229) found that in 88 spontaneous vertex deliveries, 0.5 mg of ergometrine given intravenously, compared to 10 U of oxytocin given intravenously, produced vomiting and retching in 13% of the mothers. Oxytocin did not have this side-effect. The intramuscular administration of 0.5 mg of ergometrine in 931 deliveries with delivery of the anterior shoulder was associated with a significantly higher incidence of postpartum hemorrhage and "heavy vaginal loss" when compared to syntometrine (0.5 mg ergometrine combined with 5 U oxytocin) ($p \leqslant 0.001$) (107). Intravenously administered ergometrine was also found to produce a significant elevation of diastolic blood pressure in all ten normotensive women, while syntometrine only raised the blood pressure in some of them (161). Clearly, these reports attest to the superiority of oxytocin over ergometrine for the pharmacologic management of the third stage of labor since there are fewer side-effects with the former than with the latter.

EFFECTS ON THE CERVIX

The unfavourable cervix continues to pose a problem in patients who require medically indicated induction of labor. With an unfavourable cervix, induction-delivery interval is longer and the risks of poor progress of labor or failed induction of labor are greater. With prostaglandins, rupture of the cervix and cervico-vaginal fistulae have been reported many times. Attempts have therefore been made to identify ways of improving the unfavorable cervix to a favorable one when induction of labor is necessary. The methods tried have included estrogens, intracervical balloons, oxytocin, and prostaglandins. The effect of oxytocin and prostaglandins in ripening the unfavourable cervix at or near term have been examined in two studies (361, 401). In one study 60 patients with an unfavorable cervix as determined by the Bishop score were divided into four groups, the first receiving no treatment but rest, the second received intravenous oxytocin in incremental doses up to a total of 30 U over 10 hours, the third group received 0.5 mg of PGE_2 orally every hour for 10 hours, and the last group received 1 mg PGE_2 hourly for 10 hours. All of the patients subsequently had amniotomy and intravenous oxytocin the next day if they had not yet delivered. The Bishop score appeared to be significantly better after 1 mg of PGE_2 hourly for 10 hours compared to oxytocin and controls. Also, the induction-delivery interval was markedly reduced to 7.9 hours with 1 mg of oral PGE_2 hourly priming the cervix for 10 hours. However, there were three cesarean sections, one for fetal distress, in this group compared to two with oxytocin. The Bishop score is known to be subjective and can be relied on only to a limited degree for comparison. Whether improvement in the cervix was due to a direct effect of PGE_2 or secondary to the uterine activation that it produced remained debatable. Wilson (401) compared the use of extra-amniotic PGE_2 gel, oral PGE_2 1 mg hourly for 10 hours, intravaginal PGE_2 2mg, and intravenous incremental dose oxytocin for eight hours for ripening the unfavourable cervix in 60 primigravidae. The cervix improved with all the treatments. The induction-delivery time was significantly reduced to 7.3 hours with extra-amniotic PGE_2 gel. However, unlike the study by Valentine (361), Wilson (401) had an induction-delivery interval of 13.8 hours with oxytocin, which was similar to the 12.7 hours with oral PGE_2.

Ulmsten et al. (360) showed that PGE_2 gel (0.5 mg PGE_2) was able to induce a significant ripening of the cervix compared to oxytocin. Intravenous oxytocin or intracervical PGE_2 was given randomly to 100 nulligravidae at term. If the cervix was favorable, both methods were effective in inducing labor, but when the cervix was unfavorable, significantly more patients given PGE_2 gel delivered (53%) compared to those given oxytocin (31%). Significant ripening of the cervix with a change in Bishop score from 2.9 to 6.3 occurred with PGE_2 gel compared to oxytocin.

The effect of oxytocics on the human cervix in midtrimester pregnancy was studied *in vivo* with a double, open-ended catheter (214). PGE_2, $PGF_{2\alpha}$ and oxytocin had similar but no specific effects on the intracervical canal pressure. In contrast, ergometrine caused large contractions of the cervix, usually lasting 4 to 5 minutes and occurring every 8 to 10 minutes for an hour or more. When given to patients who had previously failed to abort with therapeutic doses of prostaglandins, oxytocin produced a "reverse pattern," that is, a fall in intracervical pressure which was different from what oxytocin normally did to the cervix. Conrad and Ueland (77) extended investigations into the effect of oxytocics on the cervical tissues by using an objective index, the stretch modulus. Cervical tissue obtained from term pregnancies were tested for stiffness (stretch modulus) by elongation and by measuring the tension produced by a given stretch. It was found that the stretch moduli of the cervical tissues from patients with labor induced by PGE_2 were significantly lower than those from patients with either spontaneous or oxytocin-induced labor. Thus, in contrast to oxytocin, PGE_2 has the ability to reduce the stiffness of cervical tissue. The clinical implication is that PGE_2 may be preferable to oxytocin for the induction of labor in patients with an unfavorable cervix, but this remains to be unequivocally demonstrated in clinical trials.

■ ■ ■

Chapter 11

▪ OXYTOCIN AND THE FETUS/NEONATE

OXYTOCIN AND NEONATAL JAUNDICE

Since the initial report that oxytocin administered to the mother for the induction of labor is associated with an increased incidence of neonatal jaundice, much attention has been focused on this problem. Several explanations have been put forward to account for the increased incidence of neonatal jaundice seen with the maternal administration of oxytocin:

1) There is a higher incidence of wrong "dates" or gestational age in babies born after the elective induction of labor with oxytocin and thus jaundice of prematurity occurs. In spite of a substantial body of data indicating that it was not the incorrect gestational age which was responsible for the hyperbilirubinemia, it continued to provide discussion (249).

2) Oxytocin has an effect on bilirubin metabolism in the perinatal period, giving rise to hyperbilirubinemia.

3) Enhanced placento-fetal transfusion due to oxytocin-induced uterine contractions results in an increase in red cell mass in the neonate and therefore, neonatal hyperbilirubinemia.

Many investigations directed towards establishing the incidence and to explaining the neonatal hyperbilirubinemia associated with oxytocin administration before and during labor were reported (42, 52, 66, 67, 78, 113, 136, 180, 317, 322, 323, 344). Many of the studies showed that there was a significantly higher incidence of hyperbilirubinemia and neonatal jaundice in babies born after the oxytocin induction of labor (66, 67, 78, 113, 180, 317, 322, 323), but there were suggestions to the contrary (137). Although in one report of a small number of babies, cord bilirubin levels were found to be increased when the mothers were given oxytocin both during and for the induction of labor, other observations found that oxytocin administered during labor, as opposed to the administration of oxytocin for the induction of labor, neither affected cord bilirubin levels nor increased the incidence of hyperbilirubinemia (41, 42, 67, 78, 312, 317).

Jaundice occurred significantly faster in the oxytocin-induced group compared to the control group (322). Serum bilirubin was also significantly higher in the oxytocin group (9.7 mg/dl) compared to the control group (6.5 mg/dl).

The effect of oxytocin on jaundice was found to be independent of gestational age at birth, sex, race, epidural anesthesia, method of delivery and birth weight, all of which were known to be individually associated with neonatal jaundice (136). In another study (113), the association between oxytocin-induced labor and raised neonatal plasma bilirubin levels persisted irrespective of fetal hypoxia and acidosis at birth. Studies matched for gestational age also showed that significantly more jaundiced babies were born after oxytocin-induced labor compared to the control group (113, 317).

Dose of Oxytocin and Hyperbilirubinemia

Many reports, in which the dose of oxytocin administered to the mother was analyzed, were able to show a statistical dose-dependent oxytocin relationship with the development of neonatal jaundice (78, 113, 180, 344). However, in some studies, no demonstrable relationship could be established between the dose of oxytocin given to the mother and the severity of the jaundice (67, 322). Jeffares (180) found a highly significant association between increased oxytocin dosage in induced labors and neonatal jaundice. There was an increase in neonatal hyperbilirubinemia in mothers who had received "medium" (16 - 64 mU/min) and "high" doses of oxytocin (> 64 mU/min). This study concluded that a dose of 20 mU/min was required before the effect of oxytocin on neonatal jaundice was produced. Similarly, D'Souza et al. (113) found a highly significant correlation between cord blood bilirubin levels and the total amount of oxytocin administered to the mother in induced labor. The increase in bilirubin levels became significant after the mother had received more than 2500 mU of oxytocin. Conway et al. (78) found a significant rise in neonatal total serum bilirubin levels when the mothers received 12,000 mU oxytocin or more than 4000 mU oxytocin/kg birth weight of the infant. In the report by Buchan (52), who found increased neonatal hyperbilirubinemia in mothers with oxytocin-induced labor, the mean dose of oxytocin used was 4500 mU (range 3000 - 7000 mU). These clinical observations were substantiated by *in vitro* studies using fetal cord blood to examine the effect of varying doses of oxytocin on erythrocyte deformability (52). The erythrocyte deformability was both time- and dose-related in response to oxytocin. These observations and findings might explain the significant increase in neonatal hyperbilirubinemia associated with oxytocin used for the induction of labor but not for the augmentation of uterine contractions once labor had started. The jaundice was probably due to the longer duration of administration and the larger doses of oxytocin required in induced labor compared to the augmentation of uterine contractions during labor. A closer look at studies showing no oxytocin dose-dependent relationship in neonatal hyperbilirubinemia associated with oxytocin revealed the number of babies studied was small and sometimes included both induced as well as spontaneous labor where oxytocin was given (67, 322). Friedman et al. (136) found that oxytocin produced only a small increase in plasma bilirubin concentration of 8.6 μmol/liter (0.5 mg/dl) but did not analyze the dose of oxytocin administered to the mothers.

Pathogenesis of Oxytocin-Induced Hyperbilirubinemia

The pathogenesis of neonatal hyperbilirubinemia related to the administration of oxytocin to the mother had been studied by some investigators (52, 323). An increased osmotic fragility of the erythrocytes associated with transplacental hyponatremia was found. The cord venous blood of infants born after oxytocin-induced labor had evidence of increased hemolysis, a significantly lower packed cell volume, decreased haptoglobulin concentration, increased plasma bilirubin concentration, increased plasma lactate dehydrogenase activity, and a reduced erythrocyte deformability index and plasma osmolality when compared to the cord blood of infants born spontaneously or by cesarean section (52). *In vitro* studies using fetal cord blood showed that the changes in the erythrocyte deformability index could be induced with exogenous oxytocin and was time-related and dose-dependent in response to oxytocin. Singhi and Singh (323) found that cord serum osmolality and serum bilirubin correlated inversely with cord serum sodium, 72 hours after delivery. With a cord serum sodium level of 125 mmol/liter, serum bilirubin was 11.4 ± 2.4 mg/dl (mean ± SD) but decreased significantly to 7.7 ± 2.4 mg/dl when serum sodium increased to 135 mmol/liter. These two studies suggested that the vasopressin-like action of oxytocin resulted in the expansion of maternal extracellular fluid with dilutional hyponatremia and hypo-osmolality. Since maternal body fluids and electrolytes were in constant transplacental equilibrium, fetal hyponatremia and hypo-osmolality occurred, leading to osmotic swelling of the erythrocytes, increased deformability and hence rapid destruction with resultant hyperbilirubinemia in the neonate. Transplacental hyponatremia with accompanying hypo-osmolality had been reported when oxytocin was administered to the mother (52, 305, 324). Neonatal hyponatremia might cause convulsions in the neonatal period. Singhi and Singh (324) concluded that the transplacental passage of oxytocin administered to the mother could be responsible for the significant lowering of cord sodium osmolality found with an oxytocin infusion of greater than 16 mU/min compared to a lower dose of 2 - 16 mU/min. Certainly there is enough evidence to show that oxytocin can probably cross the human placenta (95). However, the reduction in cord serum sodium and osmolality may be compounded by the type of intravenous therapy given during labor, as shown in one retrospective study (88). Reported cases of neonatal hyponatremia always occurred in patients who had been exclusively or mainly given electrolyte-free fluids with large doses of oxytocin (305, 370). In the report by Dahlenburg (88), babies from mothers who received intravenous fluids had significantly lower serum sodium levels than those who did not, but only half of the mothers whose babies had serum sodium of less than 130 mmol/liter had received oxytocin. This suggested that although synthetic oxytocin probably played a role in hyponatremia, rapid infusions of large volumes of saline-free fluids were just as important. Schwartz and Jones (305) reported a patient who developed a *grand mal* convulsion during labor after 18 hours of intravenous oxytocin in a 5% dextrose solution. Maternal serum sodium was 117 mmol/liter and osmolality was 243 mOsmol, while cord serum sodium was 114 mmol/liter and osmolality was 243 mOsmol. The newborn had a convulsion when the serum sodium was 110 mmol/liter. While it might be suggested that intrauterine asphyxia from the maternal convulsion was responsible for the neonatal convulsion (383), it was more likely that fetal hyponatremia and hypo-osmolality were responsible since in two cases of oxytocin-induced maternal hyponatremia, neonatal convulsion occurred in one case before the mother's postpartum convulsion (370).

Thus, as recommended by most of these reports, when oxytocin is used to induce or augment labor, intravenous fluids containing saline should be used for administering the drug. However, it is noteworthy that large amounts of sodium or saline may not be ideal for the pre-eclamptic gravidas. One alternative is to use PGE_2 in such patients (315). Certainly, PGE_2-induced labors had a much lower incidence of neonatal hyperbilirubinemia compared to oxytocin induction (66, 78). A possible mechanism of the neonatal hyperbilirubinemia of oxytocin is the inhibition of hepatic bilirubin glucuronyl transferase activity by oxytocin. An *in vitro* study using serial dilutions of oxytocin compared to equal volumes of physiologic saline (control) found a significant inhibition of bilirubin glucuronyl transferase activity in the physiologic and supraphysiologic doses of oxytocin used (297). The concept of a failure to induce hepatic enzymes if labor was induced prior to its onset has been cited by many of the studies (see above). It was stated that the normal onset of labor brings about hormonal changes that were necessary for inducing fetal hepatic enzymes. Nonetheless, it was more likely that the total dose of oxytocin given to the mother and the oxytocin dose in relation to the birth weight (78) were probably more crucial.

In vitro studies showed that acetylcholine- and 5-hydroxytryptamine-induced contractions of the human umbilical artery can be enhanced by oxytocin (5 mU/ml) in a dose-dependent but reversible way (407). Only high concentrations of oxytocin (0.5 - 5.0 U/ml) by itself produced a weak response. The authors tried to explain that this umbilical artery spasm-enhancing effect of oxytocin might be contributing to the neonatal jaundice seen when mothers are given oxytocin. There is little solid experimental data to support this contention.

OTHER FETAL EFFECTS

Compared to a control group, fetal scalp temperature has been found to be significantly increased in infants whose mothers received oxytocin (23). It was suggested that the apparently normal oxytocin-induced uterine contractions might reduce uterine blood flow and thus raise the intrauterine fetal temperature. However, the reason for the elevated scalp temperature may reflect either local scalp pressure or the slowing of scalp blood flow brought about by pressure of the cervix and adjacent structures on the scalp, rather than necessarily reflecting reduced uteroplacental blood flow.

OXYTOCIN AND THE NEWBORN

A case of accidental administration of 1 ml of syntocinon (5 U oxytocin and 0.5 mg ergometrine maleate) to a 3.2 kg newborn baby was reported (397). The infant became cyanosed within 15 minutes and convulsions occurred within one hour. Ventilatory failure ensued and later, water intoxication was obvious, with weight gain, edema and a serum sodium of 102 mmol/liter. The infant recovered with treatment and was discharged at 7 days of age. Subsequent follow-up to one year of age revealed no neurologic deficit.

Several factors might contribute to the inability of the newborn animal to resist prolonged dehydration. These factors included an immature kidney with short loops of Henle and a reduced corticocapillary gradient, as well as an immature neurohypophysis which contributed to the inability to satisfactorily concentrate urine. The levels of vasopressin (AVP), oxytocin and neurophysins in the pituitary and hypothalamus, and plasma vasopressin and neurophysins of the rat during the first month of life were determined by RIA (320). Plasma AVP was less than 1.7 μU/ml in most of the animals studied. Plasma neurophysin was elevated above adult levels on day 2 and decreased with age. Oxytocin, AVP and neurophysins were present in the pituitary at brith, but at levels less than 1% of those found in adult animals. In the pituitary and hypothalamus, AVP increased rapidly in the first days after birth, but oxytocin and neurophysin remained low until 8 days and increased at 8 - 21 days of age. Pituitary AVP was 1636 ± 72/μU on the first day, and 3240 ± 228/μU on the second day, while hypothalamic AVP was 204.8 ± 18/μU on the first day and rose to 644.0 ± 60/μU on the second day. Pituitary oxytocin increased from 0.62 μU/gland on the first day to 29.3 μU on day 15, to 143.5 μU on day 30. The ratio of AVP to oxytocin was 4.4 at birth and reached unity (adult ratio) at 30 days. At birth the molar excess of AVP was largely responsible for the low molar ratio of neurophysins to oxytocin plus AVP (0.15). The ratio of neurophysins to hormones reached unity at 21 - 30 days of age. Vasotocin was not detectable in the pituitary of the rat in the first month of life. Thus, in the maturing rat, oxytocin and AVP levels changed asynchronously while neurophysin levels changed in a similar manner to those of oxytocin. It was unclear whether the low levels of pituitary oxytocin in the first few days were due to the release at parturition or continuous stimulation for oxytocin release in the postnatal period.

The same investigators looked at the response of oxytocin, AVP and neurophysin to various physiological stimuli during the first 30 days of infant life in rats (319). At all ages, the intraperitoneal injection of hypertonic saline (5%) induced a marked increase in plasma vasopressin and neurophysin up to 21 μU/ml and 51 ng/ml, respectively. In the pituitary, dehydration produced a 26 - 38% decrease in the levels of neurohypophyseal peptides with pituitary oxytocin more depleted than vasopressin. The depletion of neurohypophyseal peptides from the pituitary was greater after 24 hours of dehydration in younger rats (26%) than in older rats (7%). Plasma vasopressin was less than 1.7 μU/ml after dehydration in younger rats, but was higher in older rats. Immaturity of the neurohypophysis may contribute to the inability of newborn rats to withstand prolonged dehydration.

The neurohypophyseal hormones and their neurophysins during corticotropic maturation of the infant rats were examined using immunocytochemical localization techniques (58). Vasopressin (AVP) and neurophysin II (Np II) were present in the first magnocellular neuron population, while oxytocin and neurophysin I were present in a second neuron population. The oxytocin-Np I cellular system matured later than the AVP-Np II system since the ratio of AVP to Np II increased during the first three days after birth, but the ratio of oxytocin to Np I remained constant. In the first three weeks of life, AVP-Np II fibers in the median eminence were separated into two distinct groups: the hypothalmo-neurohypophyseal tract in the internal part of the median eminence and pericapillary endings in the external part of the median eminence. In contrast, oxytocin-Np I fibers did not demonstrate this particular distribution. The increased loading of infundibular vasopressinergic endings was related to the increased adrenocorticotropic hormone (ACTH) synthesis that has been shown to occur. Therefore, AVP and corticotropin-releasing hormone participated simultaneously in the synthesis of ACTH under physiologic stimulation of corticotropic function.

Another group of investigators found that neurophysin was detected in the rat median eminence as early as in the 18-day fetus but that vasopressin and oxytocin were detectable only after birth, at 4 and 8 days of age, respectively (385). Neurophysin and vasopressin accumulated in the palisade layer of the median eminence between four and nine days after birth. This accumulation has been postulated to be due to the relative immaturity of the capillary loops that constitute the primary plexus of the hypophyseal portal system. Oxytocin was not detected in the pituitary until one to two days after birth, in the hypothalamus after four days, and in the median eminence after eight days (69). The supraoptic nucleus matured earlier than the paraventricular nucleus.

The suprachiasmatic neurons were devoid of AVP and Np II at birth, but acquired slight AVP and Np II activity at 6 to 10 days of age, reaching adult concentrations by the third week of life (58). This appearance of vasopressin immunoreactivity in the suprachiasmatic neurons preceded the rhythmic activity of the pituitary adrenal axis and suggested a possible involvement of the suprachiasmatic vasopressin and vasopressin-like peptide in the circadian periodicity of the corticotropic function.

The hypothesis that babies born after oxytocin-induced labor had lower gastrin levels in the umbilical vein than babies born after spontaneous labor could not be confirmed or excluded in a study of gastrin levels in the umbilical arterial and venous blood of 23 babies (89). It has previously been suggested that oxytocin inhibited maternal gastrin secretion and thus reduced gastrin passage into the umbilical cord and the fetus.

■ ■ ■

Chapter 12

▪ OXYTOCIN ANALOGS

Structure-activity studies of oxytocin have permitted an understanding of the relative importance and contributions of the various amino acids of oxytocin toward its biologic activities. These studies have also enabled the synthesis of analogs which have agonist or antagonist properties. Manning and Sawyer (216) have reviewed the structure activity studies on oxytocin and vasopressin from 1954 to 1976. The discovery of some analogs have apparently been by chance while others have been from rational design or a mixture of both. Apparently no single amino acid in oxytocin is absolutely essential for the biologic activity of this molecule. While the size of the molecule may be manipulated through side-chain enlargement or reduction, optimal activities are conferred by the octapeptide structure of the native oxytocin. A ring appears to be essential but need not be a twenty-membered ring. Stereospecificity occurs since D-oxytocin is virtually inactive. The disulphide bridge *per se* is not essential for activity. Position 1 hydrogen can be replaced by an α-NH_2 group to produce deamino-oxytocin which is more potent than oxytocin itself. Position 4 of oxytocin can tolerate many substitutions without significant loss of characteristic activities.

Recent investigations have pursued the synthesis of many new analogs but attention is also focusing on the conformational structure of the oxytocin peptide backbone in relation to bioactivity and biospecificity. This chapter will review the studies on oxytocin analogs; they will be classified according to the position of the amino acid that has been substituted in the oxytocin molecule.

OXYTOCIN

Flouret et al. (132) described a method for the synthesis of oxytocin using iodine for oxidative cyclization and silica gel adsorption chromatography for purification. The product was characterized chemically and found to be biologically active.

Conformational studies of oxytocin can assist in the understanding of agonist or antagonist effects of oxytocin analogs. In one study, it was found that the oxytocin molecule has two main types of stable structures of the cyclic moeity backbone but great lability of the tail of the peptide (246). The stable structures included: 1) the β-turn conformation of the cyclic moiety backbone without closer spacing of the cyclic moiety and the tail, and; 2) structures with closely spaced N- and C-terminal parts which lack β-turn in the cyclic moiety. Tu et al. (358) used laser Raman spectroscopy to study the conformation of oxytocin. They found flexibility in the conformation of oxytocin, "β-turn"-conformation of the molecule and that the tyrosine side-chain was exposed to solvent.

A proton magnetic resonance study of conformational dynamics has examined the coordinated internal motions and chemical shifts of tocinamide (243). Such a study has provided information on the molecular conformation that could help to assist the understanding of the conformation of the oxytocin molecule. Oxytocin in aqueous solution showed no marked changes in conformation over a temperature range of 0 - 60° C (405).

A technique employing high-performance liquid chromatography to monitor oxytocin injection solutions which were prepared by dilution and aseptic filling from a concentrate of known biologic activity was recently described (221). As a result of this method, the laborious bioassays which have hitherto been used are no longer required to check the injection solution.

SUBSTITUTIONS AT POSITION 1

Khan and Sivanandaiah (186) described a method for the solid-phase synthesis of oxytocin, desamino-oxytocin and 4-thr-oxytocin. This was a rapid synthesis in the solid phase using appropriate amino acid active eaters (three equivalents) in the presence of 1-hydroxybenzotriazole (HOBT) (one equivalent) and using diphenylmethyl (DPM) to protect the thiol function of cysteine. The DPM was removed by sodium-liquid ammonia reduction. Desamino-oxytocin and 4-thr-oxytocin were synthesized using lesser amounts of amino acid active esters (1.5 equivalents) in the presence of HOBT, but requiring a lower period of coupling. The solid-phase synthesis of desamino-oxytocin using a medium containing toluene was also tried, but the toluene did not significantly accelerate the coupling rate as did the active esters.

S-Benzyl-DL-[1-^{13}C] cysteine could be prepared from $Na^{13}CN$ and used for synthesizing several oxytocin analogs (176). These analogs included [1-hemi(1-^{13}C)cystine] oxytocin, [1-hemi-D-(1-^{13}C)cystine oxytocin, [6-hemi-(1-^{13}C)cystine] -oxytocin, [6-hemi-D-(1-^{13}C)cystine] oxytocin, [1-hemi-D-(1-^{13}C)cystine, 3-D leucine] oxytocin, and [1-hemi-(1-^{13}C)-cystine, 3-D-leucine] oxytocin.

Sawyer et al. (293) synthesized several new analogs of the oxytocin antagonist [1-deaminopenicillamine] oxytocin and tested them for the ability to: 1) inhibit the uterotonic response to oxytocin in the isolated rat uterus in the absence and presence of Mg^{2+}; 2) inhibit the response to oxytocin by the rat uterus *in situ* and; 3) inhibit the response to oxytocin by the rat mammary gland *in situ*. [1-deaminopenicillamine] oxytocin has long been known to have potent antagonist activity on the isolated rat uterus suspended in Mg^{2+}-free medium, but could also act as a weak agonist on the rat uterus *in situ* and antagonize the uterotonic action of perfused oxytocin. Substituting 2-0-methyltyrosine in [1-deaminopenicillamine] oxytocin strikingly enhanced the antagonism of all uterine responses. [1-deaminopenicillamine 2-0-methyltyrosine] oxytocin and its 4-threonine analog were also potent inhibitors of the milk ejection response. Substituting 2-phenylalanine in 1-deaminopenicillamine oxytocin also enhanced the antagonist activities in all uterine assays but retained the agonist activity on milk ejection assays. Thus, changes in position 1 (1-deaminopenicillamine substitution) and position 2 (2-0-methyltyrosine or 2-phenylalanine substitution) could have additive effects on the antagonistic activities. Substitution of an 8-ornithine also enhanced the inhibitory potency *in vivo* and might therefore, also have additive effects on those of substitution in positions 1 and 2. Thus, these findings opened avenues for synthesizing newer and more potent oxytocin analogs which are effective antagonists and for evaluating their effects on the physiologic action of endogenous oxytocin. If proven effective, suitable antagonists could potentially be used for the suppression of uterine activity, as in preterm labor.

[1-deaminopenicillamine, 4-threonine] oxytocin was synthesized using protective groups by solid-phase synthesis and the [8 + 1] coupling in solution (215). This analog had no detectable agonist activity in rat vasopressin and rat uterus assays. It was a potent inhibitor of oxytocin *in vitro* with a pA_2 value of 7.46 ± 0.04 and an antivasopressor activity with a pA_2 of 6.67 ± 0.09. Therefore, substituting threonine for glutamine in the antagonist [1-deaminopencillamine] oxytocin increased the oxytocin inhibitory potency by two-fold, thus allowing [1-deaminopenicillamine,4-threonine] oxytocin to be one of the most potent inhibitors of oxytocin thus far known.

The oxytocin antagonists, [1-deaminopenicillamine,2-0-methyltyrosine] oxytocin [3,dPTyr(Me)OT] and [1-deaminopenicillamine,2-0-methyltyrosine, 4-threonine] oxytocin [4,dPTyr(Me)TOT] were synthesized and found to possess pA_2 values on the action of oxytocin on the rat uterus of 7.76 ± 0.12 and 7.6 ± 0.14, respectively, in the absence of Mg^{2+}, 7.80 ± 0.12 and 6.84 ± 0.10 *in situ* (211). [4,dPTyr(Me)TOT] was one of the most potent *in vivo* antagonists thus far reported.

Hruby et al. (175) found that the analog [1-penicillamine,2-leucine] oxytocin was a very potent competitive inhibitor of oxytocin in the oxytocic assay with a pA_2 of 7.14 and an inhibitor of oxytocin in the milk-ejection assay. It had a prolonged inhibitory effect on the uterus. Nuclear magnetic resonance studies indicated that [Pen^1, Leu^2] oxytocin and [Pen^1] oxytocin had similar conformational and dynamic properties which were different from those of the agonist, oxytocin.

Deamino-decarba-(D-D)-oxytocin and D-D-Arg-vasotocin at 10^{-4} kg/liter were shown to have an excitatory effect on the periodically oscillating neuron of the African giant snail *(Achatina fulica)* (349). Chymotrypsin treatment abolished the effect of D-D-Arg-vasotocin, but not that of deamino-decarba-(D-D)-oxytocin (348).

Proton and Carbon-13 nuclear magnetic resonance have been employed in the study of the conformational and dynamic properties of oxytocin and its analogs. The technique was used to compare oxytocin and its competititve inhibitor [1-L-penicillamine] oxytocin. Oxytocin was found to have a flexible conformation while [1-L-penicillamine] oxytocin had a more restricted one (227). These investigators felt that the results of such hormone and competitive inhibitor comparisons could provide the framework for understanding the mechanism of peptide hormone agonism and antagonism.

Nuclear magnetic resonance studies of oxytocin and the antagonist [1-penicillamine] oxytocin showed that in dimethyl sulfoxide, but not in aqueous solution, oxytocin at high concentrations (> 10 mg/ml) and [Pen^1] oxytocin in even lower concentrations, underwent some form of intermolecular interaction or association (227). However, such molecular interaction did not appear to cause any significant change in the backbone conformation of the peptides.

Conformational studies employing circular dichroism spectroscopy on three analogs of deamino-oxytocin, [1,6-aminopimelic acid] oxytocin, [1,6-deaminosuberic acid] oxytocin, and [1,6-aminoazelaic acid] oxytocin showed that their biological activities were related to their conformation and internal motions (183). [1,6-aminosuberic acid] oxytocin was the most biologically active of the three peptides, but [1,6-aminopimelic acid] oxytocin was the most rigid.

SUBSTITUTIONS AT POSITION 2

Oxytocin analogs substituted at the p- and m- positions of 2-tyrosine and/or at the N-terminal amino group are known to be inhibitors of neurohypophyseal hormones and such inhibition is usually competititve (272). The pA_2 value, which is a measure of binding of a number of substances to the oxytocin receptor, was analyzed for 1,2-substituted oxytocin analogs. The pA_2 value of these analogs suggested a significant resonance effect of p-substituted groups in 2-tyrosine when the hormone bound to its uterine receptor, whereas the N-terminal amino group exerted less clearly characterized effects. Substitution of O-methyltyrosine for tyrosine in the 2 position of the oxytocin molecule results in a loss of peak amplitude of uterine activity. However, the analog 1-desamino-1-mono-carba-[2-Try(OMe)]-oxytocin, which has an N-terminal disamination, and replacement of the 1-6 S—S bridge by CH_2S has a prolonged uterotonic effect on the porcine uterus, lasting two hours when given intravenously in doses of 5 mg/kg of body weight (81).

The uterine inhibitory effects of the analogs [2-O-ethyltyrosine] oxytocin (ethyl-OT) and [2-ethyltyrosine-8-lysine] vasopressin (ethyl-VP) were evaluated *in vitro* in nonpregnant and pregnant myometrium, and *in vivo* in the rat uterus (224). Both analogs inhibited oxytocin-induced rat uterine activity *in vivo* in a dose-dependent fashion, ethyl-LVP being more potent than ethyl-OT. In nonpregnant uteri, again both antagonists were effective in suppressing vasopressin-induced contractions, with ethyl-LVP being more potent. In pregnant myometrial strips, both antagonists were equipotent in inhibiting oxytocin-induced contractions in a dose-dependent manner at a molar ratio of antagonist to agonist of about 1.

Both [2-O-iodotyrosine] oxytocin and [2-O-methyltyrosine] oxytocin showed only weak vasopressor and antidiuretic effects, but incomplete competitive inhibition of the *in vitro* uterotonic action of oxytocin (273).

SUBSTITUTIONS AT POSITION 3

Ballardin et al. (14) used the analog [Pro^3,Gly^3]-oxytocin (PGO) in nuclear magnetic resonance studies to probe the contribution of the residues 3 and 4 which occupy the corner positions in forming a β-turn. Based on their findings and on studies from other laboratories using x-ray and/or nuclear magnetic resonance, they concluded that in peptides with type II β-turns and with L-Pro-Gly or Gly-Gly sequences in the corner positions, there was a close correlation between chemical shifts and vicinal coupling constants for glycyl residue in the second corner position. They suggested that this, in turn, can form an additional basis for the characterization of β-turns.

The third amino acid residue of oxytocin is important for its uterotonic potency. Structural conformational studies showed that removal of the phenolic hydroxyl group from its optimal position in the "active center" of oxytocin gave rise to reduced efficacy of the resulting molecule.

A comparison of the dose-response relationship of [3-phenylalanine]-oxytocin (oxypressin), with an aromatic amino acid residue in position 3, and [3-β-cyclohexylalanine] oxytocin with an aliphatic amino acid residue in position 3, in the rat uterine assay showed that oxypressin had both lower affinity for the uterine muscle receptor and decreased efficacy as compared to oxytocin (379). [3-β-cyclohexylalanine] oxytocin had an even lower affinity, but the same efficacy as oxytocin. This change in receptor affinity and efficacy was thought to be due to a reorientation of the tyrosine side-chain caused by the presence of a neighboring aromatic side-chain in position 3.

SUBSTITUTION AT POSITION 4

Threonine substitution in position 4 of oxytocin enhanced its oxytocic activity. [4-homoserine] oxytocin was synthesized using a solid-phase method and was subsequently purified. This analog had a rat oxytocic activity of 125 ± 13 U/mg and a rat antidiuretic activity of 0.24 ± 0.03 U/mg, thus giving an oxytocic antidiuretic ratio of 521, which was similar to the ratio of 512 for [4-threonine] oxytocin (359). However, [4-threonine] oxytocin had a greater oxytocic actvity of 923 ± 96 U/mg, compared to [4-homoserine] oxytocin. The analog [4-β-(2-thienyl)-L-alanine]-oxytocin has been synthesized by stepwise solution techniques and with protection of the cysteine side-chains (327). Both affinity and intrinsic activity of oxytocin was greatly reduced by the presence of thienylalanine in position 4. The intrinsic activity was even less than that for [Leu^4] oxytocin and [Phe^4] oxytocin. The analog had rat uterotonic, rat vasopressor and rat antidiuretic potencies of 0.51 ± 0.03, < 0.5 and < 0.5 U/mg, respectively. However, it showed competititve inhibition of the response to oxytocin in the avian vasopressor assay.

[Glu(OMe)^4] oxytocin and [Mpr^1,Glu(OMe)^4] oxytocin, bearing a methyl ester group in place of the carboxamide group in position 4 of oxytocin, were synthesized by Photaki et al. (268). [Glu(OMe)^4] oxytocin had potencies in the *in vitro* rat uterotonic assay of 10.5 U/mg, and an avian vasodepressor activity of 42 U/mg, while the other analog had potencies of 21.4 U/mg and 82 U/mg, respectively. Neither analog showed a response in the rat pressor assay.

The oxytocin antagonists, [1-(β-mercapto-β,β-diethylpropionic acid), 4-threonine] oxytocin (1,dEt$_2$TOT), and [1-(β-mercapto-β,β-cyclopentane-methylenepropionic acid),4-threonine] oxytocin (2,d(CH$_2$)$_5$ TOT) were prepared by Lowbridge et al. (210). The antagonists were found to have pA$_2$ values on the action of oxytocin on the rat uterus of 7.72 ± 0.11 and 7.91 ± 0.13, respectively, in the absence of Mg^{2+}, 7.36 ± 0.09 and 7.81 ± 0.09 in the presence of 0.5 nM Mg^{2+}, and 6.47 ± 0.11 and 6.94 ± 0.11 *in situ.* (2,d(CH$_2$)$_5$ TOT) was one of the most potent *in vivo* antagonists thus far developed.

Structural modifications have been made in positions 4 and 8, which make up the two β-turn corner positions of oxytocin, in order to establish their role in the activity of the molecule. The dose-response relationship of [Leu^4] oxytocin and [Phe^4] oxytocin, which are two analogs with Glu^4 side-chains in the β-turn corner position, substituted by hydrophobic and bulky groups, was compared with that of oxytocin (326). The presence of leucine or phenylalanine in position 4 caused a drastic reduction in both the affinity and the intrinsic uterotonic action of the analogs, with a greater reduction by phenylalanine than by leucine.

Using 'H nuclear magnetic resonance, a comparative analysis of [d-D-alanine] oxytocin, [4-D-alanine] oxytocin and oxytocin in dimethyl sulfoxide, indicated that the backbone amide protons of 3-alanine may be different from those of oxytocin, while those of 4-D-alanine were similar to those of oxytocin (381).

Further analysis of various physicochemical parameters suggested that these peptides can form a type II β-turn in the cyclic moiety.

SUBSTITUTION AT POSITION 5

Walter et al. (378) reported the first position 5 substitution analog of oxytocin that retained significant biologic activity. This analog, [5-aspartic acid] oxytocin was synthesized by stepwise solution techniques, employing protective groups which were eventually simultaneously removed. The purified analog had a rat uterotonic assay potency of 20.3 ± 8U/mg in the absence of

Mg^{2+} compared to 546 ± 18 U/mg for oxytocin, and an avian vasopressor potency of 41 ± 2 U/mg and a rat antidiuretic activity of 0.14 ± 0.02 U/mg, compared to 507 ± 15 and 2.7 ± 0.2 U/mg, respectively, for oxytocin. Dose-response studies on the isolated rat uterus under various experimental conditions revealed identical intrinsic activities for the analog compared to oxytocin.

[5-(N^4,N^4-dimethylasparagine)] oxytocin was synthesized and reported to have a uterotonic activity on the isolated rat uterine horn of 4.60 ± 0.03 U/mg (380). It possessed an identical intrinisic activity in the *in vitro* rat uterotonic assay as oxytocin in the presence of either 0.5 mM Ca^{2+} or at reduced Ca^{2+} levels. Position 5 residue of the oxytocin molecule had been assigned to have an "active element" responsible for the intrinsic activity of the hormone after binding to the uterine receptor.

SUBSTITUTION AT POSITION 7

The introduction of unsaturated side-chains into the proline residue 7 of oxytocin appeared to enhance the affinity of oxytocin to some of its receptors and thereby increased selectivity (231). Synthetically prepared [7-(L-3,4-dehydroproline)] oxytocin, [1-β-mercaptopropionic acid,7-(L-3,4-dehydroproline] oxytocin, and [1-L-α-hydro-β-mercaptopropionic acid,7-(L-3,4-dehydroproline)] oxytocin had increased rat uterotonic activities but had lower rat pressor activity than oxytocin. When compared to oxytocin, [7-(L-3,4-dehydroproline)] oxytocin required a smaller dose in order to produce a response of the same magnitude with a parallel shift of the uterotonic log dose versus response towards a lower concentration. This suggested that unsaturated side-chains in the proline residue in position 7 of oxytocin enhanced its binding to some of its receptors.

Position 7 of the oxytocin molecule has been suggested to be the hormone-receptor binding site (see Chapter 8). Therefore, several analogs with substitutions at position 7, [7-(3,4-dehydroproline)] oxytocin, [7-glycine]-oxytocin, and [4-threonine,7-glycine] oxytocin have been examined for their dose-response behavior compared to that of oxytocin (325). Neither the slope nor the maximal response obtained for any of the analogs tested differed significantly from the hormone. The uterotonic potencies of analogs corresponded to the relative positions along the concentration axis of their dose-response curves and to their affinities as determined by their pA_2 values. Hence, it was concluded that the differences in the uterotonic potency of these analogs resided in the differences between their affinities for the uterine receptor.

The 7-glycine substituted analogs and [8-proline] oxytocin possessed similar natriuretic activities, about 2 to 8% of the natriuretic effect of oxytocin (342). [2-Phenylalanine] oxytocin had 26% of the natriuretic effect of oxytocin, but des-9-glycinamide-oxytocin retained about 0.2% of the natriuretic action of oxytocin. The natriuretic response of these analogs appeared to be mediated by receptors which were different from the oxytocic and antidiuretic receptors for these peptides. The analog, [7-glycine] oxytocin, had markedly reduced antidiuretic and vasopressor activities, but retained considerable natriuretic activity. Stier et al. (343) wanted to determine whether the natriuretic activity of the analog [7-glycine] oxytocin was produced by decreasing proximal tubular fluid reabsorption, thus sharing a common mechanism of action with oxytocin. The analog [7-glycine] oxytocin, (10 μg/kg) was given alone or with distally-acting diuretic agents (hydrochlorothiazide) to conscious fluid-loaded rats. The analog produced diuresis and kaliuresis and the effect was greater than additive when given together with the thiazide diuretic or furosemide in terms of urine volume and sodium secretion. Triamterene blocked the kaliuresis and the effect was greater than additive when given together with the thiazide diuretic or furosemide in terms of urine volume and sodium secretion. Triamterene blocked the kaliuresis caused by [7-glycine] oxytocin, but together they had a greater than additive effect on urinary sodium excretion. The results showed that [7-glycine] oxytocin did not share a common natriuretic mechanism or site of action with the diuretics tested since an additive effect would only be seen if the analog shared a common mechanism with the diuretics. Thus, oxytocin and its analogs may increase sodium excretion – probably by decreasing next proximal tubular fluid reabsorption. The analog, [7-(thiazolidine-4-carboxylic acid)] oxytocin has been synthesized by an alternative technique employing a 6 + 3 route using hexapeptides with a performed disulfide bridge (16). The same method was subsequently adopted to synthesize the [7-(azetidine-2-carboxylic acid)] analogs of oxytocin and their biological activities were investigated (17). All activities were reduced but not to the same extent. The pressor and antidiuretic activities of the lysine-vasopressin analog were consistently found to be of a shorter duration than the corresponding activities of lysine-vasopressin. The highest biological activity of [7-azetidine-2-carboxylic acid)] oxytocin was 353 U/mg in the assay on the mammary gland *in situ,* but was greatly reduced (to 66 U/mg) on the isolated mammary strip, and to 43 U/mg on the isolated rat uterus. Both pressor and antidiuretic activities of the analog were much reduced and variable as compared to oxytocin. Thus, substitution of azetidine-2-carboxylic acid for proline at position 7 of oxytocin and lysine-vasopressin produced considerable loss of potency which varied from one biological activity to another.

SUBSTITUTIONS AT POSITIONS 4 AND 7

Lowbridge et al. (210) synthesized and analyzed the pharmacologic properties of the 4-threonine and 7-glycine substituted analogs of oxytocin. They found that [7-glycine] oxytocin had an oxytocic potency (O) of 93 ± 4 U/mg, and an antidiuretic potency (A) of 0.0056 ± 0.0003 U/mg, and therefore a high O/A ratio of 16,000. [4-threonine, 7-glycine] oxytocin had an O of 166 ± 4 U/mg and an A of 0.002 ± 0.0004 U/mg, with an O/A of 83,000. Hydroxy-$[Thr^4, Gly^7]$ oxytocin had an O of 218 ± 8 U/mg and an A of 0.004 ± 0.005 U/mg, with an O/A of 54,500. Thus, threonine substitution enhanced a greater oxytocic activity and O/A selectivity. All of the three 7-glycine-substitution analogs had a marked sensitivity to Mg^{2+} on the rat uterus assay. Because of the marked oxytocic selectivity of these peptides, they could potentially offer a greater margin of safety in clinical applications than oxytocin.

Many of the oxytocin analogs have been synthesized on the basis of structure-activity studies. Several oxytocin analogs have been synthesized which possess high oxytocic activities but negligible antidiuretic and pressor activities (341). These included [4-threonine,7-glycine] oxytocin, [1-(L-2-hydroxy-3-mercaptopropionic acid),4-threonine,7-glycine] oxytocin, and [1-(L-2-hydroxy-3-mercaptopropionic acid)] oxytocin. Stahl and Walter (341) applied structural conformation to explain some of the activities that were seen in these three analogs with a high O/A ratio. An analysis of their biologic potencies revealed that the result of chemically changing two "corner" residues of the same oxytocin analog gave rise to a linear combination of the biologic activity values of the relevant monosubstituted analogs. However, stepwise modifications of residues which were not in the corner positions, such as the introduction of the hydroxyl group in place of the amino group of the cysteinyl moiety in position 1, resulted in the nonlinear and nonspecific changes of all biologic activities. Thus, substitutions at a corner position of the two β-turns which are the important features of the oxytocin backbone structure, conferred greater additive and selective biologic effects.

SUBSTITUTION AT POSITION 8

Methionine and leucine can replace each other in natural analogs of several biologically active peptides, and it is suggested that there is a similarity between these two amino acids in their physical properties, such as solubility. An analog of oxytocin, 8-L-methionine-oxytocin, was synthesized by replacing leucine in position 8 of oxytocin (34). The analog was synthesized based on du Vigneaud's scheme for the synthesis of oxytocin. This scheme involves the coupling of the hexapeptide derivative tocinoic with the C-terminal tripeptide portion of the molecule. The reagents used for the synthesis of [8-L-methionine] oxytocin were [N-(tert-butyloxy)-carbonyl] tocinoic acid and L-prolyl-L-methionyl glycinamide. The removal of the NH_2 protecting group by acidolysis from a partially protected nonapeptide, followed by purification by chromatography, produced the oxytocin analog in a homogenous form. 8-L-methionine-oxytocin had pharmacologic activities which were characteristic of oxycin. It had 200 ± 4 U/mg of oxytocic activity determined on the isolated rat uterus, 193 ± 7 U/mg of avian depressor activity, 27 ± 0.8 U/mg of rat pressor activity, and 0.4 ± 0.2 U/mg of rat antidiuretic activity. The conformation of 8-L-methionine-oxytocin was similar to that of oxytocin. Thus, this is a potent analog of oxytocin.

[8-lysine] oxytocin can be synthesized on a solid support and was found to have an oxytocic activity of 100 ± 6 U/mmol on the isolated rat uterus bioassay (331). From the [8-lysine] oxytocin, the Σ--carbamoyl, Σ-3-carboxypropionyl and Σ-3-carboxybutyryl derivates were prepared; they had oxytocic activities of 400, 55, and 50 U/mmol, respectively. Macromolecular forms of oxytocin could be produced by coupling the [8-lysine] oxytocin through the Σ-amino group to the carboxyl groups of carboxymethylated dextrans or carboxypropionyl-gelatin. The resulting macromolecular oxytocins were water-soluble and retained significant oxytocic activity. Thus, it was suggested that [8-lysine] oxytocin could prove to be a suitable ligand for affinity chromatography of the oxytocin-binding proteins.

A hydrogen bond with the asparagine carboxamide did not appear to be necessary for the conformation of the oxytocin analog molecule to retain its hormonal action since the analogs deamino-[8-N-methylleucine] oxytocin and deamino-[8-α-hydroxyisocapriaoic acid] oxytocin retained adequate uterotonic, avian vasodepressor, antidiuretic and pressor activities (289). Neither of these analogs possessed the peptide N—H residue 8 which provided the required hydrogen.

ANALOGS – NONCOMPETITIVE INHIBITORS

Although many oxytocin inhibitors are known, almost none of them has a reversible effect since they are mainly competitive inhibitors. A number of oxytocin analogs which are noncompetititve inhibitors in the uterotonic action of oxytocin have recently been synthesized (201). The analogs were derived from deamino-oxytocin, whose disulfide bond was replaced by a thioether group and different reactive groups at the 8-carbon of the glutamic acid in position 4. The inhibitory effect of these analogs was specific for oxytocin as it was overcome by the addition of $PGF_{2\alpha}$.

THE DISULFIDE GROUP

The disulfide bridge in oxytocin has been manipulated in order to probe its contribution to the molecule and to biologically synthesize active analogs. The replacement of the disulfide bond in oxytocin by an amide group was undertaken and an analog "oxytocin lactam," [cyclo-(1-aspartic acid,6-α.β-diaminopropionic acid)] oxytocin, has been synthesized with a stepwise solution technique (329). The bridging amide bonds were formed by oxidation-reduction condensation. The analog had a rat uterotonic, avian depressor and rat antidiuretic potencies of 16 ± 2, 6.6 ± 0.6, and 5.6 ± 3.8 U/mg, respectively.

Laser Raman spectroscopy was employed to investigate the disulfide conformation in oxytocin (220). Based on the spectroscopic data of oxytocin in water and in dimethyl sulfoxide, the following conclusions were made. There was a stretching of the S–S bond and equilibrium was present among several conformations of the S–S unit of oxytocin in solution. Most of the CS–SC dihedral angles were within 30° of $+90^{\circ}$. A reinterpretation of circular dichroism spectra indicated the presence of more than one conformation for the S–S unit, which would then agree with the findings based on the Raman spectoscopy.

Laser Raman and circular dichroism spectroscopy were also used to compare the conformational properties of oxytocin agonists and antagonists (174). The agonists, oxytocin and [Gly^4] oxytocin, had very similar Raman spectra with the ν(S–S) as a singlet, while the oxytocin antagonists [Pen^1] oxytocin and [Pen^1,Leu^2] oxytocin had a singlet or a doublet ν(S–S) band. [Gly^4] oxytocin had a circular dichroism similar to oxytocin, while the circular dichroism for the antagonists [Pen^1] oxytocin and [Pen^1,Leu^2] oxytocin were similar to each other but different from that of the agonists. A further analysis revealed that the [Pen^1,Leu^2]oxytocin and [Pen^1]oxytocin molecules were less flexible than oxytocin or [Gly^4]oxytocin, thus confirming earlier observations based on nuclear magnetic resonance studies.

MISCELLANEOUS ANALOGS

Retro analogs related to oxytocin have been synthesized (235). The analogs synthesized included [D-alle^3]-retro-D-deaminotocinamide, [D-alle^3,-gly^7]-retro-D-deamino-oxytocin, [D-alle^3,gly^7,malonamide^9]-retro-D-deamino-oxytocin, and [gly^7]-retro-L-deamino-oxytocin. These retro analogs were found to be neither agonists nor antagonists in the oxytocic assay, at concentrations up to 10^{-5} M.

The oxytocin analog, 2-nitro-5-azidobenzoyl-glycyloxytocin (NAB-gly-oxytocin), was synthesized and found to be a full agonist for the stimulation of osmotic water flow and the enhancement of urea permeability in the toad urinary bladder (339).

The synthetic hydroxy analog of oxytocin, [1-(L-2-hydroxy-3-mercaptopropionic acid)]oxytocin ([L-Hmp^1]oxytocin), was found to be one and a half to two times more potent than oxytocin on the rat uterus, the rat mammary strip and the rat mammary gland *in situ,* and three times more potent on the isolated rat uterus (26). The [D-Hmp^1]oxytocin was much less potent than the [L-Hmp^1]oxytocin. The responses to oxytocin and its hydroxy analogs *in vivo* were qualitatively indistinguishable. Hydroxy analogs of oxytocin were not inactivated by pregnancy oxytocinase.

The analogs [4-(N^5,N^5-dimethylglutamine)] oxytocin and 4-(N^5,N^5-di-n-propylglutamine)] oxytocin have been synthesized and found to have potencies in the *in vitro* rat uterotonic assay of 3.0 ± 0.14 U/mg and < 0.1 U/mg, and in the avian vasodepressor assay of 4.55 ± 0.03 U/mg and < 0.1 U/mg respectively (340). There was very little rat pressor and antidiuretic activity for the second analog while the first compound produced tachyphylaxis for these activities. Oxytocinoic acid dimethylamide was found to be only 3% as potent as oxytocin in the uterotonic assay, and a marked loss of avian vasodepressor, rat pressor and antidiuretic activities was observed (355). The intrinsic activity was also greatly reduced compared to oxytocin. Thus, it appeared that replacement of the protons of the primary carboxyamide of Gly^9-NH^2 of oxytocin by methyl groups displaced the active elements of the molecule from the orientation for obtaining maximal intrinsic activity in the isolated rat uterus.

Diastereoisomers of oxytocin derivatives could be readily separated by reverse-phase high pressure liquid chromatography with good resolution (196, 197). It was suggested that this procedure offered a rapid method for screening synthetic oxytocin peptides for undesirabe diastereoisomeric by-products.

■ ■ ■

▪ BIBLIOGRAPHY

1. Aceves J. Sodium pump stimulation by oxytocin and cyclic AMP in the isolated epithelium of the frog skin. Pfluegers Arch 371:211-216, 1977.

2. Adachi A, Wilson L, Herzig N. Prostaglandin F_2 alpha, hypertonic saline, and oxytocin in midtrimester abortion. NY State J Med 77:46-49, 1977.

3. Agma A, Andersson R, Johansson C. Effect of oxytocin on sperm numbers in spontaneous rat ejaculates. Biol Reprod 18:346-349, 1978.

4. Aizawa Y, Shimizu T. Effects of estradiol and oxytocin on the release of prostaglandin-like substance from isolated rat uterus. Jpn J Pharmacol 28: 847-852, 1978.

5. Akaishi T, Negoro H. Differential effects of prostaglandin $F_{2\alpha}$ on the oxytocinergic and non-oxytocinergic neurones in the paraventricular nucleus of the lactating rat. Endocrinol Jpn 26:725-730, 1979.

6. Albasri SAI, Walker JM. Oxytocin in the urine of Brattleboro rats. J Endocrinol 81:125p-126p, 1979.

7. Alexandrova M, Soloff MS. Oxytocin receptors and parturition. I. Control of oxytocin receptor concentration in the rat myometrium at term. Endocrinology 106:730-735, 1980.

8. Alexandrova M, Soloff MS. Oxytocin receptors and parturition. II. Concentrations of receptors for oxytocin and estrogen in the gravid and non-gravid uterus at term. Endocrinology 106:736-738, 1980.

9. Alexandrova M, Soloff MS. Oxytocin receptors and parturition. III. Increases in estrogen receptor and oxytocin receptor concentrations in the rat myometrium during prostaglandin $F_{2\alpha}$-induced abortion. Endocrinology 106:739-743, 1980.

10. Amankwah K, Kaufmann RC, Roller RW, Prentice RL. A new definition of the nonstress test. Obstet Gynecol 56:48-51, 1980.

11. Aomuna S, Kohama Y, Akai K, Morita T, Nakajima S, Maeda N, Sakamoto M. Effects of oxytocin on beating properties, myosin ATPase activity and macromolecular synthesis of rat myocardial cells in cluture. Chem Pharm Bull (Tokyo) 27:1857-1863, 1979.

12. Arshat H. Extra-amniotic prostaglandin E_2 and intravenous oxytocin in termination of mid-trimester pregnancy and in the management of missed abortion and hydatidiform mole. Med J Malaysia 31:220-225, 1977.

13. Aspeslagh M-R, Vandesande F, Dierickx K. Electron microscopic immunocytochemical demonstration of separate neurophysin-vasopressinergic and neurophysin-oxytocinergic nerve fibres in the neural lobe of the rat hypophysis. Cell Tiss Res 171:31-37, 1976.

14. Ballardin A, Fischman AJ, Gibbons WA, Roy J, Schwartz IL, Smith CW, Walter R, Wyssbrod HR. Conformational studies on [Pro^3, Gly^4]-oxytocin in dimethyl sulfoxide by ^{1}H nuclear magnetic resonance spectroscopy: Evidence for a type II β turn in the cyclic moiety. Biochem 17:4443-4445, 1978.

15. Balment RJ, Brimble MJ, Forsling ML. Effect of oxytocin on renal sodium and chloride excretion in long Evans and Brattleboro rats. J Physiol (Lond) 296:89p, 1979.

16. Barber M. Jones JH. An alternative synthesis of [7-(thiazolidine-4-carboxylic acid)]-oxytocin. Int J Pept Protein Res 9:269-271, 1977.

17. Barber M, Jones JH, Stachulski AV, Bisset GW, Chowdrey HS, Hudson AL. Synthesis and biological activities of [7-(Azetidine-2-Carboxylic Acid)]-oxytocin and -lysine-vasopressin. Int J Pept Protein Res 14: 247-261, 1979.

18. Barowicz T, Styczynski H. The effect of the milking process on oxytocin activity in the blood plasma of cows during oestrous. Acta Physiol Pol 29:297-299, 1978.

19. Barrada MI, Edwards LE, Hakanson EY. Antepartum fetal testing. I. The oxytocin challenge test. Am J Obstet Gynecol 134:532-537, 1979.

20. Baskett TF, Sandy EA. The oxytocin challenge test and antepartum fetal assessment. Br J Obstet Gynaecol 84:39-43, 1977.

21. Baskett TF, Sandy EA. The oxytocin challenge test: An ominous pattern associated with severe fetal growth retardation. Obstet Gynecol 54: 365-366, 1979.

22. Baxi LV, Petrie RH, Caritis SN. Induction of labor with low-dose prostaglandin F_2 and oxytocin. Am J Obstet Gynecol 136:28-31, 1980.

23. Beck LR, Flowers Jr, CE, Blair WD. The effects of oxytocin on fetal scalp temperature. Obstet Gynecol 53:200-202, 1979.

24. Beneit JV, Hidalgo A, Tanargo JL. Effects of oxytocin on the isolated vas deferens of the rat. Br J Pharmac 69:379-382, 1980.

25. Bhakthavathsalan A, Mann LI, Tejani NA, Weiss RR. Correlation of the oxytocin challenge test with perinatal outcome. Obstet Gynecol 48:552-556, 1976.

26. Bissett GW, Chowdrey HS, Hope DB, Nilwises N, Walti M. Hydroxy analogues of oxytocin and of lysine-vasopressin. Br J Pharmacol 67:575-585, 1979.

27. Blank MS, Debias DA. Oxytocin release during vaginal distension in the goat. Biol Reprod 17:213-223, 1977.

28. Blum M, Goldman JA. Induction of abortion and labor by extra-amniotic saline, with or without addition of oxytocin, in cases of missed abortion, missed labor and antepartum foetal death. Int Surg 62:95-96, 1977.

29. Blumenstein, M, Hruby VJ. Interactions of oxytocin with bovine neurophysins I and II. Use of ^{13}C in the glycinamide-9 and half-cystine-1 positions, Biochemistry 16:5169-5177, 1977.

30. Blumenstein M, Hruby VJ, and Viswanatha V. Investigation of the interactions of oxytocin with neurophysins at low pH using carbon-13 nuclear magnetic resonance and carbon-13-labeled hormones. Biochemistry 18:3552-3556, 1979.

31. Blumenstein M, Hruby VJ, and Viswanatha V. The tyrosine ring of oxytocin undergoes hindered rotation when the hormone is bound to neurophysin. Biochem Biophys Res Commun 94:431-437, 1980.

32. Blumenstein M, Hruby VJ, and Yamamoto DM. Evidence of hydrogen-1 and carbon-13 nuclear magnetic studies that the dissociation rate of oxytocin from bovine neurophysin at neutral pH is slow. Biochemistry 17:4971-4977, 1978.

33. Blumenstein M, Hruby VJ, Yamamoto D, Yang Y. ^{13}C Nuclear magnetic resonance studies of the interactions of bovine neurophysins with (1-hemi-[1-^{13}C] cystine) oxytocin and (1-hemi-[1-^{13}] cystine, 8-arginine) vasopressin. FEBS Lett 81:347-350, 1977.

34. Bodansky M, Chandramouli N, Martinez J Walker R. Synthesis and some pharmacological properties of 8L-methionine-oxytocin. J Med Chem 22:270-273, 1979.

35. Bodanszky M, Fagan DT, Walter R, Smith CW. A ureido group containing analogue of oxytocin comprising eight amino acid residues. J Med Chem 21(3):306-308, 1978.

36. Boehm FH, Killam AP, Duncan L. The contraction stress test: a comparison of buccal and intravenous oxytocin and oral prostaglandin E_2. J Tenn Med Assoc 72:890-892, 1979.

37. Boer K, Cransberg K, Dogterom J. Effect of low-frequency stimulation of the pituitary stalk on neurohypophysial hormone release *in vivo*. Neuroendocrinology 30:313-318, 1980.

38. Bohus B, Kovacs GL, de Wied D. Oxytocin, vasopressin and memory: Opposite effects of consolidation and retrieval processes. Brain Res 157:414-417, 1978.

39. Braly P, Freeman RK. The significance of fetal heart rate reactivity

with a positive oxytocin challenge test. Obstet Gynecol 50:689-693, 1977.

40. Bonne D, Belhadj O, Cohen P. Modulation by calcium of the insulin action and of the insulin-like effect of ocytocin on isolated rat lipocytes. Eur J Biochem 75:101-105, 1977.

41. Boylan P. Oxytocin and neonatal jaundice. Br Med J 2:1004, 1976.

42. Boylan P. Oxytocin and neonatal jaundice. Br. Med J 2:504-505, 1976.

43. Braly P, Freeman RK. The significance of fetal heart rate reactivity with a positive oxytocin challenge test. Obstet Gynecol 50:689-693, 1977.

44. Breslow E, Gargiulo P. Effect of low pH on neurophysin-peptide interactions: Implications for the stability of the amino-carboxylate salt bridge. Biochemistry 16:3397-3406, 1977.

45. Briel RC, Kunz S, Kidess E. Platelet function, coagulation and fibrinolysis during termination of missed abortion and missed labor by $PGF_{2\alpha}$ and oxytocin. Acta Obstet Gynecol Scand 58:361-364, 1979.

46. Brimble MJ, Dyball REJ. Characterization of the responses of oxytocin- and vasopressin-secreting neurones in the supraoptic nucleus to osmotic stimulation. J Physiol (Lond) 271:253-271, 1977.

47. Brimble, MJ, Dyball REJ. Contrasting pattern changes in the firing of vasopressin- and oxytocin-secreting neurones during osmotic stimulation. J Physiol (Lond) 263:189P-190P, 1976.

48. Brimble MJ, Dyball REJ, Forsling ML. Oxytocin release following osmotic activation of oxytocin neurones in the paraventicular and supra-optic nuclei. J Physiol 278:69-78, 1978.

49. Brownstein MJ, Russell JT, Gainer H. Synthesis, transport and release of posterior pituitary hormones. Science 207:373-378, 1980.

50. Brzezinska-Slebodzinsak E, Adamczyk J. Partial purification and characterization of the oxytocin-inactivating enzymes from chicken liver. Acta Biochim Pol 26:407-416, 1979.

51. Buchan P. Maternal hormone profile in oxytocin induced labor. Br J Obstet Gynaecol 86:861-865, 1979.

52. Buchan P. Pathogenesis of neonatal hyperbilirubinaemia after induction of labor with oxytocin. Br Med J 2:1255-1257, 1979.

53. Buijs RM. Intra- and Extrahypothalamic vasopressin and oxytocin pathways in the rat. Cell Tissue Res 192:423-435, 1978.

54. Buijs RM. Immunocytochemical demonstration of vasopressin and oxytocin in the rat brain by light and electron microscopy. J Histochem Cytochem 28:357-360, 1980.

55. Buijs RM, Pevet P. Vasopressin- and oxytocin-containing fibres in the

pineal gland and subcommissural organ of the rat. Cell Tissue Res 205:11-17, 1980.

56. Buijs RM, Swaab DF. Immuno-electron microscopical demonstration of vasopressin and oxytocin synapses in the limbic system of the rat. Cell Tissue Res 204:355-365, 1979.

57. Buijs RM, Swaab DF, Dogterom J, van Leeuwen FW. Intra- and extra-hypothalamic vasopressin and oxytocin pathways in the rat. Cell Tissue Res 186:423-433, 1978.

58. Burlet A, Chateau M, Czernichow P. Immunocytochemical study of neurohypophysial peptides during corticotrophic maturation of infant rats. Cell Tissue Res 101:315-325, 1979.

59. Burlet A, Chateau M, Czernichow P. Infundibular localization of vasopressin, oxytocin and neurophysins in the rat; its relationship with corticotrope function. Brain Res 168:275-286, 1979.

60. Caffrey MH, Nett TM, Kozlowski GP. *In vitro* analysis of the participation of oxytocin and vasopressin in the gonadotropin releasing hormone-induced release of LH. Proc Soc Exp Biol Med 159:444-448, 1978.

61. Carsten ME. Calcium accummulation by human uterine microsomal preparations: Effects of progesterone and oxytocin. Am J Obstet Gynecol 133:598-601, 1979.

62. Carsten ME, Miller JD. Effects of Prostaglandins and oxytocin on calcium release from a uterine microsomal fraction. J Biol Chem 252:1576-1581, 1977.

63. Celis ME, Vivas A, Orlwoski RC, Walter R. Neurohypophyseal hormone fragment-induced release of MSH. Brain Res 149:516-518, 1978.

64. Chanley WA, Bagoyo MM, Bryant-Greenwood GD. *In Vitro* response of relaxin-treated rat uterus to prostaglandins and oxytocin. Prostaglandins 14:763-769, 1977.

65. Chan WY. Relationship between the uterotonic action of oxytocin and prostaglandins: Oxytocin action and release of PG-activity in isolated non-pregnant and pregnant rat uteri. Biol Reprod 17:541-548, 1977.

66. Chew WC. Neonatal hyperbilirubinaemia: A comparison between prostaglandin E_2 and oxytocin induction. Br Med J 2:679-680, 1979.

67. Chew WC, Swann IL. Influence of simultaneous low amniotomy and oxytocin infusion and other maternal factors on neonatal jaundice: a prospective study. Br Med J 1:72-73, 1977.

68. Choy VJ, Watkins WB. Immunocytochemical study of the hypothalamo-neurohypophyseal system. II Distribution of neurophysin, vasopressin and oxytocin in the normal and osmotically stimulated rat. Cell Tiss Res 180:

467–490, 1977.

69. Choy VJ, Watkins WB. Maturation of hypothalamo-neurohypophyseal system. I. Localization of neurophysin, oxytocin and vasopressin in the hypothalamus and neural lobe of the developing rat brain. Cell Tiss Res 197:325-336, 1979.

70. Clarke G, Fall CHD, Lincoln DW, Merrick LP. Effects of cholinoceptor antagonists on the suckling-induced and experimentally evoked release of oxytocin. Br J Pharmacol 63:519-527, 1978.

71. Clarke G, Herring LP, Lincoln DW. Two separate cholinergic mechanisms for regulation of oxytocin release. Br J Pharmacol 61:123P, 1977.

72. Clarke G, Lincoln DW, Merrick LP. Dopaminergic control of oxytocin release in lactating rats. J Endocrinol 83:409-420, 1979.

73. Clarke G, Lincoln DW, Schoot PVD. Neuropharmacological studies on the central inhibition of oxytocin release. Br J Pharmacol 62:424p-425p, 1978.

74. Clarke G, Merrick LP. A tentative identification of the synaptic transmitters involved in the neural regulation of oxytocin release. J Physiol (Lond) 277:19P-20P, 1978.

75. Clarke G, Wood P, Merrick L, Lincoln DW. Opiate inhibition of peptide release from the neurohumoral terminals of hypothalamic neurones. Nature 282:746-748, 1979.

76. Cohen P, Nicolas P, and Camier M. Biochemical aspects of neurosecretion: Neurophysin-neurohypophyseal hormone complexes. Curr Top Cell Regul 15:263-318, 1979.

77. Conrad JT, Ueland K. The stretch modulus of human cervical tissue in spontaneous, oxytocin-induced and prostaglandin E_2-induced labor. Am J Obstet Gynecol 133:11-14, 1979.

78. Conway DI, Read MD, Bauer C, Martin RH. Neonatal jaundice – A comparison between intravenous oxytocin and oral prostaglandin E_2. J Int Med Res 4:241-246, 1976.

79. Convert O, Griffin JH, DiBello C, Nicolas P, Cohen P. Selectively enriched ^{13}C-labeled proline-7 and ^{13}C-labeled leucine-8 ocytocins as probes for the analysis of peptide hormone interactions with bovine neurophysin I using ^{13}C nuclear magnetic resonance spectoscopy. Biochem 16:5060-5065, 1977.

80. Corey JF. Elective induction of labour. Med J Aust 2:755, 1977.

81. Cort N, Einarrson S, and Viring S. Actions of oxytocin and long-acting carba oxytocin analog on the porcine myometrium *in vitro* and *in vivo*. Am J Vet Res 40:430-432, 1979.

82. Costrini R, Barbieri RCC, Costa G, Biandri M, Galloro P, Cattani L,

Maselli G, Damiani G, Delli Colli G, Innamorati G. Disappearance of oxytocin-induced uterine tiredness by treatment with fructose-1, 6-disphosphate. Experimental evidence. Int J Vit Nutr Res 48:412-417, 1978.

83. Crankshaw DJ, Branda LA, Matlib MA, Daniel EE. Localization of the oxytocin receptor in the plasma membrane of rat myometrium. Eur J Biochem 86:481-486, 1978.

84. Crowley WR, Donohue TL, George JM, Jacobowitz DM. Changes in pituitary oxytocin and vasopressin during the estrous cycle and after ovarian hormones: evidence for mediation by norepinephrine. Life Sci 23:2579-2586, 1978.

85. Crowley WR, George JM, Jacobowitz DM. Levels of arginine-vasopressin and oxytocin in the pituitary gland and individual hypothalamic nuclei of the genetically obese Zucker rat. J Endocrinol 77:417-418, 1978.

86. Csaba G, Ronai A, Laszlo V, Zsuzsanna D, Berzetei I. Amplification of hormone receptors by neonatal oxytocin and vasopressin treatment. Horm Metab Res 12:28-31, 1980.

87. Curet LB, Olson RW. Oxytocin challenge test and urinary estriols in the mangement of high-risk pregnancies. Obstet Gynecol 55:296-300, 1980.

88. Dahlenburg GW, Burnell RH, Braybrook R. The relationship between cord serum sodium levels in newborn infants and maternal intravenous therapy during labor. Br J Obstet Gynaecol 87:519-522, 1980.

89. Davidson DC, Rogers IM, Ardill J, Buchanan KD. Neonatal gastric hyperactivity. Further analysis of oxytocin effect. Arch Dis Child 50:818-820, 1975.

90. Dawood MY. Oxytocin, new data may help establish the ideal dose. Contemp Obstet Gynecol 13:181-191,1979.

91. Dawood MY, Lauersen NH, Trivedi D, Ylikorkala O, Fuchs F. Studies on oxytocin in the baboon during pregnancy and delivery. Acta Endocrinol 91:704-718, 1979.

92. Dawood MY, Raghavan KS, Pociask C. Radioimmunoassay of oxytocin. J Endocrinol 76:261-270, 1978.

93. Dawood MY, Raghavan KS, Pociask C, Fuchs F. Oxytocin in human pregnancy and parturition. Obstet Gynecol 51:138-143, 1978.

94. Dawood MY, Raghavan KS, Ylikorkala O. Oxytocin i blodet under graviditet og fodsel. Ugeskr Laeger 140:2992-2995, 1978.

95. Dawood MY, Wang CF, Gupta R, Fuchs F. Fetal contribution to oxytocin in human labor. Obstet Gynecol 52:205-209, 1978.

96. Dawood MY, Ylikorkala O, Trivedi D, Fuchs F. Oxytocin in maternal circulation and amniotic fluid during pregnancy. J Clin Endocrinol Metab

49:429-434, 1979.

97. Dawood MY, Ylikorkala O, Trivedi D, Gupta R. Oxytocin levels and disappearance rate and plasma follicle-stimulating hormone and luteinizing hormone after oxytocin infusion in men. J Clin Endocrinol Metab 50:397-400, 1980.

98. Delaney RL, Dunn AJ, Tintner R. Behavioral responses to intracerebroventrically administered neurohypophyseal peptides in mice. Horm Behav 11:348-362, 1978.

99. del Pozo E, Kleinstein J, Brundel Re R, Derrer F, Martin-Perez J. Failure of oxytocin and lysine-vasopressin to stimulate prolactin release in humans. Horm Metab Res 12:26-28, 1980.

100. Deslauriers R, Smith ICP, Stahl GL, Walter R. Studies of the interaction of bovine neurophysin-II with [1-hemi-D(3-^{13}C)-cystine] oxytocin and [1-hemi-(3^{13}C)cystine] oxytocin. Int J Pept Protein Res 13:78-87, 1979.

101. de Wied D. Neuropeptides: Effects on motivation, learning and memory processes. pp 15-24, in *Clinical Psychoneuro-endocrinology in Reproduction,* eds. L Carenza, P Pancheri, L Zichella, Academic Press, New York, 1979.

102. de Wied D, Bohus B, Van Ree JM, Urban I, Greidanus TBVW. Neurohypophyseal hormones and behaviour. pp 201-210, in *Neurohypophysis,* eds AM Moses, L Share, Karger, Basel 1977.

103. Dierickx K, Vandesande F. Immunocytochemical localization of vasopressinergic and oxytocinergic neurons in the human hypothalamus. Cell Tissue Res 184:15-27, 1977.

104. Dierickx K, Vandesande F. Immunocytochemical neurons in the human hypothalamus. Cell Tissue Res 196:203-212, 1979.

105. Dierker Jr. LJ. Oxytocin challenge test. Am J Obstet Gynecol 132: 702-703, 1978.

106. Dittman R, Belcher J. False-negative oxytocin challenge test. N Eng J Med 298:56, 1978.

107. Djahanbakhch O, Vere M, Garner NHN, Morris ED. The intra-muscular use of oxytocic agents for prophylactic management of the third stage of labour. Br J Clin Pract 32:137-138, 1978.

108. Dogterom J, Greidanus TBVW, Swaab DF. Evidence for the release of vasopressin and oxytocin into cerebrospinal fluid: Measurements in plasma and CSF of intact and hypophysectomized rat. Neuroendocrinology 24:108-118, 1977.

109. Dogterom J, Snijdewint FGM, Pavet P, Buijs RM. On the presence of neuropeptides in the mammalian pineal and subcommissural organ. Prog Brain Res 52:465-470, 1979.

110. Dogterom J, Snijdewint FGM, Pavet P, Swaab DF. Studies on the presence of vasopressin, oxytocin and vasotocin in the pineal gland, subcommissural organ and fetal pituitary gland: failure to demonstrate vasotocin in mammals. J Endocrinol 84:115-123, 1980.

111. Dominguez AN, Corbeira MSC, Gonzales-Conde CS, Montes JM, Herandez EG. Significance of the transient bradycardic pattern in prepathological oxytocin stress test. Gerburtshilfe Perinatol 182:294, 1978.

112. Druzin, ML, Paul RH, Gratacos J. Current status of the contraction stress test. J Reprod Med 23:222-226, 1979.

113. D'Souza SW, Black P, MacFarlane T, Richards B. The effect of oxytocin in induced labor on neonatal jaundice. Br J Obstet Gynaecol 86:133-138, 1979.

114. Dubin NH, Ghodgaonkar RB, King TM. Role of prostaglandin production in spontaneous and oxytocin induced uterine contractile activity in *in vivo* pregnant rat uteri. Endocrinology 105:47-51, 1979.

115. Dyball FE, Shaw FD. Inactivation of oxytocin release during prolonged electrical stimulation of the rat neural lobe. J Physiol (Lond) 284:123p, 1978.

116. Dyball REJ, Nordmann JJ. Reactivation of veratridine of hormone release from the K^+ depolarized rat neurohypophysis. J Physiol (Lond) 269: 65P-66P, 1977.

117. Dyball REJ, Wakerley JB, Poulain DA, Brimble MJ. Electro-physiological determinants of neurosecretion. pp 78-87, in: *Neurohypophysis,* eds. AM Moses, L Share, Karger, Basel, 1977.

118. Dyer RG. Electrochemical deposition of iron onto rat oxytocinergic neurones does not result in milk-ejection. J Endocrinol 81:65-73, 1979.

119. Eggers TR, Fliegner JR. Water intoxication and syntocinon infusion. Aust NZJ Obstet Gynaecol 19:59-60, 1979.

120. Egley C, Suzuki K. Intrauterine fetal demise after negative oxytocin challenge test. Obstet Gynecol 50:545-575, 1977.

121. Eiler H, Sims M. Mastitis-Metritis-Agalactia complex in sows: Effect of the dosage of oxytocin in intramammary pressure in lactating healthy sows. Am J Vet Res 40:1100-1103, 1979.

122. Ellendorff F, Forsling ML, Parvizi N, Smidt D, Taverne M, Williams H. Prostaglandin $F_{2\alpha}$ adminstration and oxytocin release in the pig and miniature pig. Br J Pharmacol 62:412P, 1978.

123. Ellendorff F, Forsling M, Parvizi N, Williams H, Taverne M, Smidt D. Plasma oxytocin and vasopressin concentrations in response to prostaglandin injection into the pig. J Reprod Fertil 56:573-577, 1979.

124. Ellendorff F, Poulain DA, Vincent JD. An electrophysiological study of septal input to oxytocin and vasopressin neurones in the supraoptic nucleus of the rat. J Physiol (Lond) 284:124P, 1978.

125. Erker EF, Chan WY. The site and the mechanism of phenoxybenzamine potentiation of the pressor response to oxytocin and vasopressin: *in vivo* and isolated aortic strips studies. J Pharmacol Exp Ther 202:287-293, 1977.

126. Falconer L, Mitchell MD, Mountford LA, Robinson JS. Plasma oxytocin concentrations during the menstrual cycle in the rhesus monkey, macaca mulatta. J Reprod Fertil 59:69-72, 1980.

127. Farahani G, Fenton AN. Fetal heart rate acceleration in relation to the oxytocin challenge test. Obstet Gynecol 49:163-166, 1977.

128. Fernstrom JD, Fisher LA, Cusack BM, Gillis MA. Radioimmunologic detection and measurement of nonapeptides in the pineal gland. Endocrinology 106:243-251, 1980.

129. Ferrier BM. Action of oxytocin on mammary gland: influence of hypotonicity. Can J Physiol Pharmacol 57:314-317, 1979.

130. Flint APF, Forsling ML, Mitchell MD. Blockade of the Ferguson Reflex by lumbar epidural anaesthesia in the parturient sheep: Effects on oxytocin secretion and uterine venous prostaglandin F levels. Horm Metab Res 10:545-547, 1978.

131. Flint APF, Mitchell MD, Sheldrick EL. Delayed luteal regression in ewes immunized against oxytocin. J Physiol (Long) Nov. 296:85p-86p, 1979.

132. Flouret G, Ferada S, Kato T, Gualtieri R, Lipkowski A. Synthesis of oxytocin using iodine for oxidative cyclization and silica gel adsorption chromatography for purification. Int J Pept Protein Res 13:137-141, 1979.

133. Flouret G, Terada S, Yang F, Nakagawa SH, Nakahara T, Hechter O. Iodinated neurohypophyseal hormones as potential ligands for receptor binding and intermediates in synthesis of tritiated hormones. Biochemistry 16:2119-2124, 1977.

134. Forsling ML, Taverne MAM, Parvizi N, Elsaesser F, Smidt D, Ellendorf F. Plasma oxytocin and steroid concentrations during late pregnancy, parturition and lactation in the miniature pig. J Endocrinol 82:61-69, 1979.

135. Fribourg S. Vaginal evacuation. Obstet Gynecol 51:740, 1978.

136. Friedman L, Lewis PJ, Clifton P, Bulpitt CJ. Factors influencing the incidence of neonatal jaundice. Br Med J 1:1235-1237, 1978.

137. Friedman L, Lewis P, Harvey D. Oxytocin and neonatal jaundice. Br Med J 2:1233, 1977.

138. Friedman EA, Sachtleben MR. Effect of oxytocin and oral prostaglandin E_2 on uterine contractility and fetal heart rate patterns. Am J Obstet

Gynecol 130:403-407, 1978.

139. Fuchs F, Fuchs A-R, Dawood MY. Neuroendocrine factors in human labor. pp 357-367, in *Clinical Psychoneuroendocrinology in Reproduction,* eds. L Carenza, P Pancheri, L Zichella, Academic Press, New York, 1979.

140. Gainer H, Sarne Y, Brownstein MJ. Biosynthesis and axonal transport of rat neurohypophysial proteins and peptides. J Cell Biol 73:365-381, 1977.

141. Gal D, Neuhoff S, Lilling MI, Tancer ML. False negative oxytocin challenge test: Report of three cases. Am J Obstet Gynecol 133:111-113, 1979.

142. Galli-Gallardo SM, Pang PKT, Oguro C. Renal responses of the Chilean toad calyptocephalella candiverbera, and the mud puppy, necturus maculosus, to mesotocin. Gen Comp Endocrinol 37:134-136, 1976.

143. Garite TJ, Freeman RK, Hochleutner I, Luizey EM. Oxytocin challenge test. Obstet Gynecol 51:614-618, 1978.

144. Garrioch DB. The effect of indomethacin on spontaneous activity in the isolated human myometrium and on the response to oxytocin and prostaglandin. Br J Obstet Gynaecol 85:47-52, 1978.

145. Gazis D. Plasma half-lives of vasopressin and oxytocin analogs after IV injection in rats. Proc Soc Exp Biol Med 158:663-665, 1978.

146. Gazis D, Roy J, Schwartz IL. Prolonged responses by the rat uterus *in vivo* to deamino-oxytocin disproportionate to its estimated plasma half-life. Endocrinology 106:805-810, 1980.

147. George JM. Immunoreactive vasopressin and oxytocin: Concentration in individual human hypothalamic nuclei. Science 200:342-343, 1978.

148. George JM, Forrest J. Vasopressin and oxytocin content of microdissected hypothalamic areas in rats with hereditary diabetes insipidus. Neuroendocrinology 21:275-279, 1976.

149. Gershkovich ZI. Mechanism of action of neurohypophyseal hormones on sodium excretion and diuresis. Neurosci Behav Physiol 7:9-12, 1976.

150. Gillespie SL, Pickford M. A search for the factors responsible for absence of recovery of the normal vascular response to oxytocin following sympathetic nerve injury. Q J Exp Physiol 61:331-339, 1976.

151. Glatz TH, Weitzman RE, Nathanielz PW, Fisher DA. Metabolic clearance rate and transplacental passage of oxytocin in the pregnant ewe and fetus. Endocrinology 106:1006-1011, 1980.

152. Golbus MS, Creasy RK. Uterine priming with oral prostaglandin E_2 prior to elective induction with oxytocin. Prostaglandins 14:577-581, 1977.

153. Goossens N, Dierickx K, Vandesande F. Immunocytochemical

localization of vasotocin and mesotocin in the hypothalamus of lacertilian reptiles. Cell Tissue Res 200:223-227, 1979.

154. Goossens N, Dierickx K, Vandesande F. Immunocytochemical localization of vasotocin in isotocin in the preopticohypophysial neurosecretory system of teleosts. Gen Comp Endocrinol 32:371-375, 1977.

155. Gorewit RC. Method of determining oxytocin concentrations in unextracted sera; characterization of lactating cattle (403904). Proc Soc Exp Biol Med 160:80-87, 1979.

156. Griffin JH, DiBello C, Alazard R, Nicolas P, Cohen P. Carbon–13 nuclear magnetic resonance studies of the binding of selectively ^{13}C-enriched oxytocins to the neurohypophyseal protein, bovine neurophysin II. Biochemistry 16:4194-4198, 1977.

157. Grimes DA, Cates Jr, W, Petitli DB, Parker J. Fatal uterine rupture during oxytocin-augmented, saline-induced abortion. Am J Obstet Gynecol 130:591-592, 1978.

158. Grosvenor CE, Mena F. Alterations in the oxytocin-induced intramammary pressure response after mechanical stimulation of the mammary gland of the anesthetized lactating rat. Endocrinology 104:443-447, 1979.

159. Guzek JW, Swiderska Z, Motylewska A. The adrenergic influences on the hypothalamic and neurohypophysial oxytocic activity in long term dehydrated male white rats. Acta Physiol Pol 29:221-229, 1978.

160. Hachamovitch B, Bracken MB, Simons H. Saline-instillation abortion with laminaria and megadose oxytocin. Am J Obstet Gynecol 135:327-330, 1979.

161. Hacker NF, Biggs JSG. Blood pressure changes when uterine stimulations are used after normal delivery. Br J Obstet Gynaecol 86:633-636, 1979.

162. Haldar J, Sawyer WH. Inhibition of oxytocin release by morphine and its analogs. Proc Soc Exp Biol Med 157:476-480, 1978.

163. Hamilton MG, Hirst M. Antagonism of smooth muscle responses to oxytocin and vasopressin by salsolinol. Eur J Pharmacol 44:279-281, 1977.

164. Hanew K, Masataka S, Rennels EG. Effect of indoles, AVT, oxytocin and AVP on prolactin secretion in rat pituitary clonal (2B8) cells (40858). Proc Soc Exp Biol Med 164:257-261, 1980.

165. Hashmonai M, Torem S, Argov S, Barzilai A, Schramek A. Prolonged post-vagotomy gastric atony treated by oxytocin. Br J Surg 66:550-551, 1979.

166. Hayward JN. Hypothalamic magnocellular neuroendocrine cell activity and neurohypophysial hormone release. pp 67-77, in *Neurohypophysis*, eds. AM Moses, L Share, Karger Basel, 1977.

167. Hems DA, Rodrigues LM, Whitton PD. Rapid stimulation by vasopressin, oxytocin and angiotension II of glycogen degradation in hepatocyte suspensions. Biochem J 172:311-317, 1978.

168. Hillman RB, Ganjam VK. Hormonal changes in the mare and foal associated with oxytocin induction of parturition. J Reprod Fertil (Suppl) 27:541-546, 1979.

169. Hochman J, Weiss G, Steinetz BG, O'Byrne EM. Serum relaxin concentrations in prostaglandin- and oxytocin-induced labor in women. Am J Obstet Gynecol 130:473-474, 1978.

170. Hoffman PL. The influence of arginine-vasopressin and oxytocin on ethanol dependence and tolerance. Curr Alcohol 5:5-16, 1978.

171. Homburg R, Insler V. Loss of beat-to-beat variability and negative oxytocin challenge test: an ominous prognostic sign. Int J Gynaecol Obstet 17:159-163, 1979.

172. Horowitz AJ. Midtrimester abortion utilizing intra-amniotic prostaglandin $F_{2\alpha}$, laminaria and oxytocin. J Reprod Med 21:236-240, 1978.

173. Huddleston JF, Sutliff G, Carney Jr., FE, Flowers Jr., CE. Oxytocin challenge test for antepartum fetal assessment. Report of a clinical experience. Am J Obstet Gynecol 135:609-614, 1979.

174. Hruby VJ, Deb KK, Fox J, Bjarnason J, Tu AT. Conformational studies of peptide hormones using laser Raman and circular dichroism spectroscopy. A comparative study of oxytocin agonists and antagonists. J Biol Chem 253:6060-6067, 1978.

175. Hruby VJ, Deb KK, Yamamoto DM, Hadley ME, Chan WY. [1-penicillamine, 2-leucine] oxytocin. Synthesis and pharmacological and conformational studies of a potent peptide hormone inhibitor. J Med Chem 22:7-12, 1979.

176. Hruby VJ, Viswanatha V, Yany YCS. Synthesis of S-benzyl-DL-[1-^{13}C] cysteine and its incorporation into oxytocin and [8-arginine]-vasopressin and related compounds by total synthesis. Separation of disastereoisomers by partition chromatography and HPLC. J Label Comp Radiopharm 17:801-811, 1980.

177. Ishikawa M, Fuchs A-R. Effects of epinephrine and oxytocin on the release of prostaglandin F from the rat uterus *in vitro*. Prostaglandins 15:89-101, 1978.

178. Jackson R, George JM. Oxytocin in microdissected hypothalamic nuclei significant differences between prepubertal and sexually mature female rats. Neuroendocrinology 31:158-160, 1980.

179. Jarell SE, Sokol RJ. Clinical use of stressed and nonstressed monitoring techniques. Clin Obstet Gynecol 22:617-632, 1979.

180. Jeffares MJ. A multifactorial survey of neonatal jaundice. Br J Obstet Gynaecol 84:452-455, 1977.

181. Jofre ME, Green R, Gillespie A. Antepartum stress cardiotocography using buccal oxytocin. Aust NZ J Obstet Gynaecol 18:230-233, 1978.

182. Kaplan E. A generalized epileptiform convulsion after intra-amniotic prostaglandin with intravenous oxytocin infusion. A case report. S Afr Med J 53:27-29, 1978.

183. Kawano K, Kobayashi Y, Kyogoku Y, Morikawa T, Sakakibara S. Conformational studies of oxytocin analogues with different ring sizes. I. Circular dichroism. Biochem Biophys Acta 578:87-95, 1979.

184. Kelly J, Swanson LW. Additional forebrain regions projecting the posterior pituitary: preoptic region, red nucleus of the stria terminals, and zona incerta. Brain Res 197:1-9, 1980.

185. Kennedy TG. Effect of oxytocin on estrogen-induced uterine luminal fluid accumulation in ovariectomized rats and the role of prostaglandins. Can J Physiol Pharmacol 56:908-914, 1978.

186. Khan SA, Sivanandaiah KM. Solid-phase synthesis of oxytocin desamino-oxytocin and their 4-Thr-oxytocin using active esters in presence of 1-hydroxybenzotriazole. Int J Pept Protein Res 12:164-169, 1978.

187. Kierse MJNC, Sokolewicz JJ, Frankena A, Jaszman L. Comparison or oral prostaglandin E_2 and intravenous oxytocin for induction of labor in hypertensive pregnancies. Eur J Obstet Gynecol Reprod Biol 10:231-237, 1980.

188. Kirchberger MA, Hechter O, Walter R, Schwartz IL. Neurohypophyseal peptide action on adenylate cyclase and hydro-osmotic properties of toad urinary bladder epithelium. Biochim Biophys Acta 500:246-255, 1977.

189. Knox GE, Huddleston JF, Flowers Jr., CE, Eubanks A, Sutliff G. Management of prolonged pregnancy: Results of a prospective randomized trial. Am J Obstet Gynecol 134:376-384, 1979.

190. Koller WS, Curet LB. Fetal activity determinations and oxytocin challenge test for assessment of fetal well-being. Obstet Gynecol 52:176-178, 1978.

191. Kovacs GL, Bohus B, Versteeg DHG, DeKloet R, De Wied D. Effect of oxytocin and vasopressin on memory consolidation: Sites of action and catecholaminergic correlates after local microinjection into limbic-midbrain structures. Brain Res 175:303-314, 1979.

192. Kovacs GL, Vecsei L, Telegdy G. Opposite action of oxytocin to vasopressin in passive avoidance behaviour in rats. Physiol Behav 20:801-802, 1978.

193. Krisch B. Immunohistochemical and electron microscopic study of the rat hypothalamic nuclei and cell clusters under various experimental conditions. Possible sites of hormone release. Cell Tiss Res 174:109-127, 1976.

194. Kumarasamy T. Recognition of extrauterine pregnancy. Am J Obstet Gynecol 135:549, 1979.

195. Kuriyama H, Suzuki H. Effects of prostaglandin E_2 and oxytocin on the electrical activity of hormone-treated and pregnant rat myometria. J Physiol (Lond) 260:335-349, 1976.

196. Larsen B, Fox BL, Burke MF, Hruby VJ. The separation of peptide hormone diastereoisomers by reverse phase high pressure liquid chromatography. Int J Pept Protein Res 13:21-21, 1979.

197. Larsen B, Viswanatha V, Chang SY, Hruby VJ. Reverse phase high pressure liquid chromatography for the separation of peptide hormone diastereoisomers. J Chromatogr Sci 16:207-210, 1978.

198. Lassof S, Altura B. Do pial terminal arterioles respond to local perivascular application of the neurohypophyseal peptide hormones, vasopressin and oxytocin? Brain Res 196:266-269, 1980.

199. Laudanski T, Akerlund M, Batra S. Differences in the effects of vasopressin and oxytocin on rabbit myometrial activity and a possible mediation of prostaglandins. J Reprod Fertil 51:355-361, 1977.

200. Leake RD, Weitzman RE. Developmental pharmacokinetics of the posterior pituitary hormones. Clin Perinatol 6:65-85, 1979.

201. Lebl M, Bojanovska V, Barth T, Jost K. Oxytocin challenge analogs effective as noncompetitive inhibitors in uterotonic test. Experientia 34:1543-1545, 1978.

202. Leodolter S, Muller-Tyl E, Philipp K, Salem G, Janisch H. The influence of ritodrine-HCl, Bunitrolol and oxytocin on uteroplacental perfusion. Z Geburtshilfe Perinatol 183:111-117, 1979.

203. Levin M, Kim YN. Oxytocin challenge test at Maryland General Hospital. Md State Med J 29:61-63, 1980.

204. Li JH, de Sousa RC. Effects of Ag^+ on frog skin: interactions with oxytocin, amiloride and ouabain. Experientia 33:433-436, 1976.

205. Lin CC, Moawad AH, River P, Pishotta FT. An OCT-reactivity classification to predict fetal outcome. Obstet Gynecol 56:17-23, 1980.

206. Lincoln DW, Clarke G, Mason CA, Dreifuss JJ. Physiological mechanism determining the release of oxytocin in milk ejection and labor. pp 101-109, in *Neurohypophysis,* eds. AM Moses L Share, Karger, Basel, 1977.

207. Lindmark G, Nilsson BA. A comparative study of uterine activity in labour induced with prostaglandin $F_{2\alpha}$ or oxytocin and in spontaneous

labour. I Pattern of the uterine contractions. Acta Obstet Gynecol Scand 55: 453-460, 1976.

208. Lindmark G, Nilsson BA. A comparative study of uterine activity in labor induced with prostaglandin $F_{2\alpha}$ or oxytocin and in spontaneous labour. II Characteristics of uterine activity and twin effect on the progress of labour. Acta Obstet Gynecol Scand 56:87-94, 1977.

209. Lorenz RP, Pagano JS. A case of intrauterine fetal death after a negative oxytocin challenge test. Am J Obstet Gynecol 130:232-233, 1978.

210. Lowbridge J, Manning M, Haldar J, Sawyer WH. Synthesis and some pharmacological properties of [4-threonine, 7-glycine] oxytocin, [1-(L-2hydroxy-3-mercaptopropionic acid), 4-threonine, 7-glycine] oxytocin, (hydroxy [Thr^4, Gly^7] oxytocin), and [7-glycine] oxytocin, peptides with high oxytocic antidiuretic selectivity. J Med Chem 20:120-123, 1977.

211. Lowbridge J, Manning M, Seto J, Haldar J, Sawyer WH. Synthetic antagonists of *in vivo* responses by the rat uterus to oxytocin. J Med Chem 22:565-569, 1979.

212. Lykkesfeldt G, Osler M. A comparison of three methods for inducing labor, oral prostaglandin E_2, buccal desamino-oxytocin, intravenous oxytocin. Acta Obstet Gynecol Scand 58:321-325, 1979.

213. Macdonald AA, Forsling ML, Williams H, Ellendorf F. Plasma vasopressin and oxytocin concentration in the conscious pig foetus: responses to hemorrhage. J Endocrinol 81:124P-125P, 1979.

214. Mackenzie IZ. The effect of oxytocics on the human cervix during midtrimester pregnancy. Br J Obstet Gynaecol 83:780-785, 1976.

215. Manning M, Lowbridge J, Seto J, Haldar J, Sawyer WH. [1-Deaminopenicillamine, 4-threonine] oxytocin, a potent inhibitor of oxytocin. J Med Chem 21:179-182, 1978.

216. Manning M, Sawyer WH. Structure-activity studies on oxytocin and vasopressin 1954-1976. From empiricism to design. pp 9-21, in *Neurohypophysis,* eds. AM Moses, L Share, Karger, Basel, 1977.

217. Marcum RG. False negative oxytocin challenge test. Am J Obstet Gynecol 127:894, 1977.

218. Marklee HV, Warr JL, Branda LA. Oxytocin binding sites in lactating rabbit mammary gland. Can J Biochem 56:968-976, 1978.

219. Marsden DE, Correy JF. The relative value of the non-stress and stress cardiotocograph: a protocol and pilot study. Aust NZJ Obstet Gynaecol 19:86-90, 1979.

220. Maxfield FR, Scheraga HA. A Raman spectroscopic investigation of the disulfide conformation in oxytocin and lysine vasopressin. Biochemistry 16:4443-4449, 1977.

221. Maxl F, Krummen K. Dilution control of aseptically filled oxytocin ampoules, using high performance liquid chromatography. Pharm Acta Helv 53:207- 210, 1978.

222. McCranie WM, Niebyl JR. A false negative oxytocin challenge tests. Obstet Gynecol 49:241-243, 1977.

223. McKenna P, Shaw RW. Hyponatremic fits in oxytocin-augmented labors. Int J Gynaecol Obstet 17:250-252, 1979.

224. Melin P, Akerlund M, Vilhardt H. Antagonism of the myometrial response to oxytocin and vasopressin by synthetic analogues. Dan Med Bull 26:126-128, 1979.

225. Mena F, Pacheco P, Martinez G, Grosvenor CE. Reflex regulation of autonomic influence upon the oxytocin-induced contractile response of the mammary gland in the anesthetized rat. Endocrinology 104:751-756, 1979.

226. Meraldi J-P, Hruby VJ. Studies on the molecular association of oxytocin and related compounds in dimethyl sulfoxide. J Am Chem Soc 98:6408-6410, 1976.

227. Meraldi J-P, Hruby VJ, Brewster AIR. Relative conformational rigidity in oxytocin and [1-penicillamine] oxytocin: a proposal for the relationship of conformational flexibility to peptide hormone agonism and antagonism. Proc Natl Acad Sci 74:1373-1377, 1977.

228. Mitchell MD, Mountford LA, Natale R, Robinson JS. Concentrations of oxytocin in the plasma and amniotic fluid of rhesus monkeys (macaca mulatta) during the later half of pregnancy. J Endocrinol 84:473-478, 1980.

229. Moir DD, Amoa AB. Ergometrine or oxytocin? Blood loss and side-effects at spontaneous vertex delivery. Br J Anaesth 51:113-117, 1979.

230. Monaco ME, Kidwell WR, Lippman ME. Neurohypophysial-hormone-response cell line derived from a dimethylbenzanthracene-induced rat mammary tumor. Biochem J 188:437-441, 1980.

231. Moore S, Felix AM, Meienhofer J, Smith CW, Walter R. Pharmacological effects of introducing a double bond into a binding site of oxytocin. Analogues with L-3, 4-dehydroproline in position 7. J Med Chem 20:495-500, 1977.

232. Moos F, Richard P. The inhibitory role of β-noradrenergic receptors in oxytocin release during suckling. Brain Res 169:595-599, 1979.

233. Morgan DB, Kirwan NA, Hancock KW, Robinson D, Howe JG, Ahmad S. Water intoxication and oxytocin infusion. Br J Obstet Gynaecol 84:6-12, 1977.

234. Morris JF, Sokol HW, Valtin H. One neuron-one hormone? Recent evidence from brattleboro rats, pp 58-66, in *Neurohypophysis,* eds. AM Moses, L. Share, Karger, Basel,1977.

235. Nakagawa SH, Yand F, Kato T, Fouret G, Hechter O. Retro analogs related to oxytocin. Int J Pept Protein Res 8:465-479, 1976.

236. Neely DP, Stabenfeldt GH, Sauter CL. The effect of exogenous oxytocin on luteal function in mares. J Reprod Fertil 55:303-308, 1979.

237. Nelson GH, Bryans Jr. CI. A comparison of oral prostaglandin E_2 and intravenous oxytocin for induction of labour in normal and high risk pregnancies. Am J Obstet Gynecol 126:549-554, 1976.

238. Neuhoff SD, Gal D, Tancer ML. False negative oxytocin challenge test result. NY State J Med 10:1537-1540, 1979.

239. New restrictions on oxytocin use. FDA Drug Bull 8:30, 1978.

240. Newcomb R, Booth WD, Rowson LEA. The effect of oxytocin treatment on the levels of prostaglandin F in the blood of heifers. J Reprod Fertil 49:17-24, 1977.

241. Newton M. The role of the oxytocin reflexes in three interpersonal reproductive acts: coitus, birth and breast feeding. pp 411-418, in *Clinical Psychoneuroendocrinology in Reproduction,* eds L Carenza, Pancheri P, Zichella L, Academic Press, New York, 1979.

242. Ng KH, Wong WP. Risk of haemorrhage in oxytocin stress test. Lancet 2:698, 1976.

243. Nicholls LJF, Jones CR, Gibbons WA. Proton magnetic resonance study of conformational dynamics coordinated internal motions, and chemical shifts of tocinamide. Biochemistry, 16:2248-2254, 1977.

244. Nicolas P, Dessen P, Camier M, Cohen P. Cooperative binding of oxytocin to bovine neurophysin. FEBS Lett 86:188-192, 1978.

245. Nicolas P, Wolff J, Camier M, DiBello C, Cohen P. Importance of neurophysin dimer and of tyrosine-49 in the binding of neurohypophyseal peptides. J Biol Chem 253:2633-2639, 1978.

246. Nikiforovich GV, Leonova VI, Galaktionov, Chipens GI. Theorectical conformational analysis of oxytocin molecule. Int J Pept Protein Res 13:363-373, 1979.

247. Nissenson R, Flouet G, Hechter O. Opposing effects of estradiol and progesterone on oxytocin receptors in rabbit uterus. Proc Natl Acad Sci 75:2044-2048, 1978.

248. Nissenson RA, Flouret G, Hechter O. Oxytocin receptors coupled to uterine contraction in estrogen-dominated rabbits. Biochim Biophys Acta 628:209-219, 1980.

249. Noble AD. Oxytocin and neonatal jaundice. Br Med J 1:239, 1978.

250. Nordmann JJ, Dyball REJ. Effects of veratridine on Ca fluxes and the release of oxytocin and vasopressin from the isolated rat neurophypophysis, J Gen Physiol 72:297-304, 1978.

251. Novo-Dominguez A, Sandoval C, Botella-Llusia J. L'epreuve de tolerance foetale a l'oxytocine antepartum. J Gynecol Obstet Reprod (Paris) 7:819-826, 1978.

252. Odendaal HJ. Hyperstimulation of the uterus during the oxytocin stress test. Obstet Gynecol 51:380-383, 1978.

253. Olund A, Larson B. Comparison of extra-amniotic instillation of rivanol and $PGF_{2\alpha}$ either separately or in combination followed by oxytocin for second trimester abortion. Acat Obstet Gynecol Scand 57:333-336, 1978.

254. Orr JW, Huddleston JF, Knox GE, Goldenberg RL, Davis RO. False negative oxytocin challenge test associated with abdominal pregnancy. Am J Obstet Gynecol 133:108-110, 1979.

255. Ottesen B, Ulrichsen H, Wagner G, Fahrenkrug J. Vasoactive intestinal polypeptide (VIP) oxytocin induced activity of the rabbit myometrium. Acta Physiol Scand 107:285-287, 1979.

256. Pang PKT, Sawyer WH. Renal and vascular responses of the bullfrog (Rana catebeiana) to mesotocin. Am J Physiol 235:F151-F155, 1978.

257. Parer JT, Afonso JF. Validity of the weekly interval between oxytocin challenge test. Am J Obstet Gynecol 127:204-205, 1977.

258. Parish DC, Pickering BT. Differential effects of colchicine on transport in the oxytocin and vasopressin-containing neurones of the hypothalamo-neurohypopyhsial system in the rat. J Endocrinol 79(2):24P-25P, 1978.

259. Parisi M, Chevalier J, Bourguet J. Influence of mucosal and serosal pH on antidiuretic action in frog urinary bladder. Am J Physiol 237:F483-F489, 1979.

260. Parvez S, Raza-Bukhari A, Ismahan G. The influence of hypophysectomy, adrenalectomy, progesterone, oxytocin and estradiol on the metabolic fate of ^{3}H-epinephrine in central and peripheral regions during late pregnancy. Am J Obstet Gynecol 134:13-19 1979.

261. Paulin C, Czernichow P, Dubois MP. Immunocytological evidence for oxytocin neurones in the human fetal hypothalamus. Cell Tissue Res 188:259-264, 1978.

262. Pavlou C, Barker GH, Roberts A, Chamberlain GVP. Pulsed oxytocin infusion in the induction of labour. Br J Obstet Gynaecol 85:96-100, 1978.

263. Pearlmutter AF, McMains C. Interaction of bovine neurophysin with oxytocin and vasopressin measured by temperature jump relaxation. Biochemistry 16:628-633, 1977.

264. Pearlmutter AF, Soloff MS. Characterization of the metal ion requirement for oxytocin-receptor interaction in rat mammary gland membranes. J Biol Chem 254:3899-3906, 1979.

265. Peck TM. Physicians' subjectivity in evaluating oxytocin challenge test. Obstet Gynecol 56:13-16, 1980.

266. Pedersen CA, Prange AJ Jr. Induction of maternal behaviour in virgin rats after intracerebroventricular administration of oxytocin. Proc Natl Acad Sci 76:6661-6665, 1979.

267. Perry G, Siegal B, Held B. Uterine trauma associated with midtrimester abortion induced by intra-amniotic prostaglandin $F_{2\alpha}$ with and without concomitant use of oxytocin. Prostaglandins 13:1147-1159, 1977.

268. Peterson DR, Oparil S, Flouret G, Carone FA. Handling of angiotensin II and oxytocin by renal tubular segments perfused *in vitro*. Am J Physiol 232:F319-F324, 1977.

269. Photaki I, Tzougraki C, Kotsira-Eugonopoulos C. Synthesis and some pharmacological properties of [Glu $(oMe)^4$]. Int J Pept Protein Res 13: 426-433, 1979.

270. Pickford GE, Strecker EL. The spawning reflex response of the Kellifish, *Fundulus heteroclitus:* isotocin is relatively inactive in comparison with arginine vasotocin. Gen Com Endocrinol 32:132-137, 1977.

271. Platt JE, LiCause MJ. Effects of oxytocin in larval ambystoma tigrinum: Acceleration of induced metamorphosis and inhibition of the antimetamorphic. Gen Comp Endocrinol 41:84-91, 1980.

272. Pliska V. Affinity of 1,2-substituted oxytocin analogues to the uterus receptor: Free-Wilson and Hansch analysis. Experientia 34:1190-1192, 1978.

273. Pliska V, Marbach P, Vasak J, Rudinger J. [2-0-Iodotyrosine]-oxytocin: Basic pharmacology and comments on their potential use in binding sites. Experientia 33:367-369, 1977.

274. Pliska V, Rudinger J. Modes of inactivation of neurohypophysial hormones: Significance of plasma disappearance rate for their physiological responses. Clin Endocrinol 5:735-845, 1975.

275. Polain DA, Wakerley JB, Dyball REJ. Electrophysiological differentiation of oxytocin and vasopressin-secreting neurones. Proc R Soc Lond 196:367-384, 1977.

276. Pranchev N. Determination of oxytocin sensitivity-the sensitive test of Smyth, by a monitor follow-up of cases with prolonged pregnancy with or without drugs. Akush Ginekol (Sofia) 17:244-247, 1978.

277. Pratila MG, Pratilas V. Pulmonary edema occuring during emergence from anesthesia. Mt Sinai J Med 46:333-334, 1979.

278. Pratt D, Diamont F, Yen H, Bieniarz J, Burd L. Fetal stress and nonstress tests: An analysis and comparison of their ability to identify fetal

outcome. Obstet Gynecol 54:419-423, 1979.

279. Propping D, Stubblefield PG, Golub J, Zuckerman J. Uterine rupture following midtrimester abortion by laminaria, prostaglandin $F_{2\alpha}$ and oxytocin: Report of two cases. Am J Obstet Gynecol 128:689-690, 1977.

280. Raghavan KS, Singh J, Chhabra JK. Development of a radioimmunoassay for oxytocin. Indian J Med Res 66:787-793, 1977.

281. Reaves TA Jr., Hayward JN. Immunocytochemical identification of vasopressinergic and oxytocinergic neurones in the hypothalamus of the cat. Cell Tissue Res 196:117-122, 1979.

282. Roberts JS, McCracken JA. Does prostaglandin $F_{2\alpha}$ released from the uterus by oxytocin mediate the oxytocic action of oxytocin? Biol Reprod 15:457-463. 1976.

283. Roberts JS, McCracken JA, Gavagan JE, Soloff MS. Oxytocin-stimulated release of prostaglandin $F_{2\alpha}$ from ovine endometrium *in vitro:* Correlation with estrus cycle and oxytocin-receptor binding. Endocrinology 99:1107-1114, 1976.

284. Robinson ICAF, Walker JM. Extraction of small amounts of oxytocin from biological fluids by means of agarose-bound neurophysin. J Endocrinol 80:191-202, 1979.

285. Roca RA, Garofalo EG, Martino I, Piriz H, Rieppi G, Maraffi M, Ohahian C, Gadola L. Effects of oxytocin antiserum and of indomethacin on hCG-induced ovulation in the rabbit. Biol Reprod 19:552-557, 1978.

286. Roca RA, Garofalo EG, Piriz H, Martino I, Rieppi G. Influence of the estrous cycle on the action of oxytocin on rat ovarian contractility *in vivo.* Fertil Steril 28:205-208, 1977.

287. Roca R, Garofalo EG, Piriz H, Martino I, Rieppi G, Sala M. Effects of oxytocin on *in vitro* ovarian contractility during the estrous cycle of the rat. Biol Reprod 15:464-466, 1976.

288. Roux JF, Mofid M, Moss PL, Dmytus KC. Effect of elective induction of labor with prostaglandins $F_{2\alpha}$ and E_2 and oxytocin on uterine contraction and relaxation. Am J Obstet Gynecol 127:718-722, 1977.

289. Roy J, Roy U, Gazis D, Schwartz IL. Biofunctional evaluation of the hydrogen bond linking the ring and tail β-turns of oxytocin. Proc Natl Acad Sci 76:3309-3313, 1979.

290. Sakamoto H, Den K, Yamamoto K, Arai T, Kwai S, Oyama Y, Yoshida T, Takagi S. Study of oxytocin receptor in human moyometrium using highly specific ^{3}H-labeled oxytocin. Endocrinology 26:515-522, 1979.

291. Salerno NJ, Kay TR. A further challenge to the validity of the weekly interval between oxytocin challenge tests. Am J Obstet Gynecol 130: 849-851, 1978.

292. Sandstrom B, Solheim F, Magnusson S. The effect of oxytocin on hypertonic saline abortion. Acta Obstet Gynecol Scand [Suppl] 66:129-131, 1977.

293. Sawyer WH, Gazis JHD, Szeto J, Bankowski K, Lowbridge J, Turan A, Manning M. The design of effective *in vivo* antagonists of rat uterus and milk ejection responses to oxytocin. Endocrinology 106:81-91, 1980.

294. Scanlon JW, Suzuki K, Shea E, Tromick E. Clinical and neurobehavioural effects of repeated intrauterine exposure to oxytocin: A prospective study. Am J Obstet Gynecol 132:294-296, 1978.

295. Scanlon JW, Suzuki K, Shea E, Tronick E. A prospective study of the oxytocin challenge test and newborn neurobehavioural outcome. Obstet Gynecol 54:6-11, 1979.

296. Schams D, Schmidt-Polex B, Kruse V. Oxytocin determination by radioimmunoassay in cattle. Acta Endocrinol 92:258-270, 1979.

297. Schikler KN, Cohen MI, McNamara H. Oxytocin administration. Am J Dis Child 130:1377, 1976.

298. Schlageter N, Janis RA, Gualtieri RT, Hechter O. Effects of oxytocin and methacholine on cyclic nucleotide levels of rabbit myometrium. Can J Physiol Pharmacol 58:243-248, 1980.

299. Schmidt WK, Holaday JW, Loh HH, Way EL. Failure of vasopressin and oxytocin to antagonize acute morphine antinociception or facilitate narcotic tolerance development. Life Sci 23:151-158, 1978.

300. Schorderet M, Gross A, De Sousa RC. Cyclic AMP levels in isolated frog skin epithelium: effects of phosphodiesterase inhibitors, oxytocin and catecholamines. Mol Cell Endocrinol 11:105-116, 1978.

301. Schriefer JA, Lewis PR, Miller JW. Effect of dopamine on length of gestation and on the release of fetal oxytocin in rats. J Pharmacol Exp Ther 212:431-434, 1980.

302. Schroeder BT, Chakaraborty J, Soloff MS. Binding of [^{3}H] oxytocin to cells isolated from the mammary gland of the lactating rat. J Cell Biol 74:428-440, 1977.

303. Schubert F, George JM, Rao MR. Monosodium glutamate (MSG) alters protein, vasopressin and oxytocin in microdissected hypothalamic areas in newborn and adult rats. Life Sci 26:650-656, 1980.

304. Schulman H, Lin CC, Saldana L, Randolph G. Quantitative analysis in the oxytocin challenge test. Am J Obstet Gynecol 129:239-244, 1977.

305. Schwartz RH, Jones RWA. Transplacental hyponatraemia due to oxytocin. Br Med J 1:152-153, 1978.

306. Secher NJ, Arnsbo P, Wallin L. Hemodynamic effects of oxytocin

(syntocinon) and methyl ergometrine (methergin) on the systemic and pulmonary circulations of pregnant anaesthetized women. Acta Obstet Gynecol Scand 57:97-103, 1978.

307. Seitchik J, Chatkoff ML. Oxytocin-induced uterine hypercontractility pressure wave forms. Obstet Gynecol 48:436-441, 1976.

308. Seitchik J, Chatkoff ML, Hayashi RH. Intrauterine pressure waveform characteristics of spontaneous and oxytocin- or prostaglandin $F_{2\alpha}$-induced active labor. Am J Obstet Gynecol 127:223-227, 1977.

309. Sellers SM, Hodgson HT, Mitchell MD, Anderson ABM, Turnbull AC. Release of prostaglandin after amniotomy is not mediated by oxytocin. Br J Obstet Gynaecol 87:43-46-1980.

310. Sengupta Sur S, Rabbani LD, Libman L, Breslow E. Fluorescence studies of native and modified neurophysins. Effects of peptides and pH. Biochemistry 18:1026-1036, 1979.

311. Sethi S. Oxytocin challenge test. Am J Nurs 78:2112-2115, 1978.

312. Seto J, Haldar J, Sawyer WH, [1-deaminopenicillamine, 4-threonine] oxytocin, a potent inhibitor of oxytocin. J Med Chem 21(2):179-181, 1978.

313. Seybold VS, Miller JW, Lewis PR. Investigation of dopaminergic mechanism for regulating oxytocin release. J Pharmacol Exp Ther 207:605-610, 1978.

314. Sheldrick EL, Mitchell MD, Flint APF. Delayed luteal regression in ewes immunized against oxytocin. J Reprod Fertil 59:37-42, 1980.

315. Silverstone PI. Transplacental hyponatraemia due to oxytocin. Br Med J 1:362, 1978.

316. Simmons WH, Walter RH. Carboxyamidopeptidase: Purification and characterization of a neurohypophyseal hormone inactivating peptide from toad skin. Biochemistry 19:39-48, 1980.

317. Sims DG. Oxytocin and neonatal jaundice. Br Med J 2:1070, 1976.

318. Sims MH, Eiler H. Porcine mastitis-metritis-agalactia (MMA) syndrome: Mammary gland responsiveness to oxytocin given to healthy sow during lactation. Am J Vet Res 40:1104-1106, 1979.

319. Sinding C, Robinson AG, Seif SM. Levels of neurohypophyseal peptides in the rat during the first month of life. II. Response to physiological stimuli. Endocrinology 107:755-760, 1980.

320. Sinding C, Seif SM, Robinson AG. Levels of neurohypophyseal peptides in the rat during the first month of life. I. Basal levels in plasma, pituitary and hypothalamus. Endocrinology 107:749-754, 1980.

321. Singh PJ, Hofer MA. Oxytocin reinstates maternal olfactory cues for nipple orientation and attachment in rat pups. Physiol Behav 20:385-389, 1978.

322. Singh M, Singhi S. Oxytocin infusion during labour and neonatal jaundice. Indian Pediatr 15:399-402, 1978.

323. Singhi S, Singh M. Pathogenesis of oxytocin-induced neonatal hyperbilirubinaemia. Arch Dis Child 54:400-402, 1979.

324. Singhi S, Singh M. Transplacental asymptomatic hyponatremia following oxytocin infusion during labour. Indian J Med Res 70:55-57, 1979.

325. Smith CW, Botos CR, Skala G, Walter R. Conformation-activity studies of oxytocin. Effects of structural modification at corner positions of the β-turns on the uterotonic activity. J Med Chem 20(11):1457-1460, 1977.

326. Smith CW, Chan S, Walter R. Dose-response behaviour on the isolated rat uterus of oxytocin analogs with modifications at binding sites. J Pharmacol Exp Ther 203:120-124, 1977.

327. Smith CW, Skala G, Walter R. Synthesis and some pharmacological properties of [4-β-(2-Thienyl)-L-alanine] oxytocin. J Med Chem 21(1):115-117, 1978.

328. Smith WG, Allen HH, Collins JA, Roth J. Induction of midtrimester abortion with intra-amniotic area and intravenous oxytocin. Am J Obstet Gynecol 127:228-231, 1977.

329. Smith WS, Walter R. Replacement of the disulfide bond in oxytocin by an amide group. Synthesis and some biological properties of [cyclo-(1-L-aspartic acid, 6-L-α,β-diaminopropionic acid)] oxytocin. J Med Chem 21(1): 117-120, 1978.

330. Smythies JR, Beaton JM, Bemington F, Bradley RJ, Morin RF. On the molecular mechanism of interaction between the neurophysins and oxytocin and vasopressin. J Theor Biol 63:33-48, 1976.

331. Snell CR, Smyth DG. Biologically active macromolecular forms of oxytocin. Biochem J 165:43-47, 1977.

332. Sofroniew MV. Projections from vasopressin, oxytocin, and neurophysin neurones to neural targets in the rat and human. J Histochem Cytochem 28:475-478, 1980.

333. Soloff MS. Regulation of oxytocin action at the receptor level. Life Sci 25:1453-1460, 1979.

334. Soloff MS. Uterine receptors for oxytocin: correlation between antagonist potency and receptor binding. Br J Pharmacol 57:381-386, 1976.

335. Soloff M, Alexandrova M, Fernstrom MJ. Oxytocin receptors: triggers for parturition and lactation. Science 204:1313-1314, 1979.

336. Soloff MS, Schroeder BT, Chakraborty J,, Pearlmutter AF. Characterization of oxytocin receptors in the uterus and mammary gland. Fed Proc 36:1861-1866, 1977.

337. Sorbe B. Active pharmacologic management of the third stage of labour: a comparison of oxytocin and ergometrine. Obstet Gynecol 52:649-697, 1978.

338. Spellacy WN, Cruz AC, Kalra PS, Kellner KR, Quinlan RW, Buhi WC, Birk SA. Oxytocin challenge test results compared with simultaneously studied serum human placental lactogen and free estriol levels in high-risk pregnant women. Am J Obstet Gynecol 135:917-923, 1979.

339. Stadel JM, Goodman DBP, Galardy RE, Rasmussen H. Synthesis and characterization of 2-nitro-5-azidobenzoyl glycyloxytocin, an oxytocin photo-affinity label. Biochemistry 17:1403-1408, 1978.

340. Stahl G, Smith GW, Walter R, Tsegenidis T, Stavropoulos G, Cordopatis P, Theodoropoulos D. Oxytocin and lysine-vasopressin with N^5 N^5, dialkylglutamine in the 4 position: Effect of introducing sterically hindered groups into the hydrophilic cluster neurohypophyseal hormones. J Med Chem 23:213-217, 1980.

341. Stahl GL, Walter R. Combined high oxytocic with negligible anti-diuretic and pressor activities in multisubstituted oxytocins. J Med Chem 20: 492-495, 1977.

342. Stier Jr., CT, Manning M, Sawyer WH. Effects of structural changes on the natriuretic activity of oxytocin analogs in conscious rats. Proc Soc Exp Biol Med 164:167-172, 1980.

343. Stier CT, Manning M, Sawyer WH. Natriuretic effect of [7-glycine] oxytocin in the presence of diuretic agents in conscious rats. J Pharmacol Exp Ther 212:412-417, 1980.

344. Stockman R, Sidiropoulos D, von Muralt G. Perinatal influence on hyperbilirubinemia of the newborn infant. J Pediat 90:858-860, 1977.

345. Strauss JH, Wilson M, Cadwell D, Otterson W, Martin AO. Laminaria use of midtrimester abortions induced by intra-amniotic prostaglandins $F_{2\alpha}$ with urea and intravenous oxytocin. Am J Obstet Gynecol 134:260-264, 1979.

346. Swanson LW, McKellar S. The distribution of oxytocin- and neurophysin-stained fibers in the spinal cord of the rat and monkey. Comp Neur 188:87-106, 1979.

347. Takahashi K, Diamond F, Bieniarz J, Yen H, Burd L. Uterine contractility and oxytocin sensitivity in preterm, term and post-term pregnancy. Am J Obstet Gynecol 136:774-779, 1980.

348. Takeuchi H, Matsumoto M, Mori A. Modification of effects of biological active peptides, caused by enzyme treatment, on the excitability of identifiable giant neurones of an African giant snail (Achatina fulica ferussac). Experientia 33:249-251, 1977.

349. Takeuchi H, Sakai A, Mori A. Effects of three synthetic peptides analogous to neurohypophyseal hormones on the excitability of giant neurones of *Achatina fulica* ferussac. Experientia 32:1554-1556, 1976.

350. Taverne M, Ellendorf F. Pattern of oxytocin release during labour in the miniature pig. J Endocrinol 75:42P-43P, 1977.

351. Taverne MAM, Naaktgeboren C, Elsaesser F, Forsling ML, vander Wyeden GC, Ellendorf F, Smidt D. Myometrial electrical activity and plasma concentrations of progesterone, estrogens, oxytocin during late pregnancy and parturition in the miniature pig. Biol Reprod 21:1125-1134, 1979.

352. Telegdy G, Kovacs GL. Role of monomines .n mediating the action of hormones on learning and memory. pp 249-268, in *Brain Mechanisms in Memory and Learning*, ed Brazier MAB, Raven Press, New York, 1979.

353. Tepperman HM, Beydoun SN, Abdul-Karim RW. Drugs affecting myometrial contractility in pregnancy. Clin Obstet Gynecol 20:423-445.

354. Theodosis DT, Burlet C, Bondier J-L, Dreifuss JJ. Morphology of membrane changes during neurohypophyseal hormone release in a hibernating rodent. Brain Res 154:371-376, 1978.

355. Ting YF, Smith CW, Stahl GL, Walter R. Effect of changing the COOH-terminal amide group present in the hydrophilic cluster of oxytocin to dimethylamide. J Med Chem 23:693-695, 1980.

356. Toaff ME, Hezroni J, Toaff R. Induction of labour by pharmacological and physiological doses of intravenous oxytocin. Br J Obstet Gynaecol 85:101-108, 1978.

357. Tribollet E, Clarke G, Dreifuss JJ, Lincoln DW. The role of central adrenergic receptors in the reflex release of oxytocin. Brain Res 142:69-84, 1978.

358. Tu AT, Bjarnason JB, Hruby VJ. Conformation of oxytocin studied by laser Raman spectroscopy. Biochim Biophys Acta 533:530-533, 1978.

359. Turan A, Manning M, Haldar J, Sawyer WH. Synthesis and some pharmacological properties of [4-homoserine] oxytocin. J Med Chem 20:1169-1172, 1977.

360. Ulmsten U, Wingerup L, Andersson K-E. Comparison of prostaglandin E_2 and intravenous oxytocin for induction of labour. Obstet Gynecol. 54: 581-584, 1979.

361. Valentine BH. Intravenous oxytocin and oral prostaglandin for repening of the unfavourable cervix. Br J Obstet Gynaecol 84:846-854, 1977.

362. van den Bergh AS, Haspels AA. Termination of second trimester pregnancy with intra-amniotic administration of 16-phenoxy-20-tetranor-PGE_2-methylsulfonamide (SHB286) alone and combined with oxytocin and calcium. Contraception 18:635-639, 1978.

363. van Leeuwen FW. Immunocytochemical specifity for peptides with special reference to arginine-vasopressin and oxytocin. J Histochem Cytochem 28:479-482, 1980.

364. van Leeuwen FW, deRaay C, Swaab DF, Fisser B. The localization of oxytocin, vasopressin, somatostatin and luteinizing hormone releasing hormone in the rat neurohypophysis. Cell Tissue Res 202:189-201, 1979.

365. van Leeuwen FW, Swaab DF. Specific immunoelectronmicroscopic localization of vasopressin and oxytocin in the neurohypophysis of the rat. Cell Tissue Res 177:493-501, 1977.

366. Van Vossel-Doeninck J, Dierickx K, Van Vossel, A, Vandesande F. Electron microscopic immunocytochemical demonstration of separate vasotocinergic nerve fibers in the median eminence of the frog hypophysis. Cell Tissue Res 204:29-36, 1979.

367. Vandesande F, Dierickx K. The activated hypothalamic magnocellular neurosecretory system and the one neurone – one neurohypophysial hormone concept. Cell Tissue Res 200:29-33, 1979.

368. Vasicka A, Kumaresan P, Han GS, Kumaresan M. Plasma oxytocin in initiation of labour. Am J Obstet Gynecol 130:263-273, 1978.

369. Vasicka A, Kumaresan P, Li C, Scheffs J. Immunologic and biological activity of oxytocin in midtrimester pregnancy. Am J Obstet Gynecol 127:171-175, 1977.

370. Vere MF, Sellers SM. Transplacental hyponatremia due to oxytocin. Br Med J 1:362, 1978.

371. Veznik Z, Holub A, Zraly Z, Kummer V, Holcak V, Jost K, Cort JH. Regulation of bovine labour with a long-acting carba-analog of oxytocin. A preliminary report. Am J Vet Res 40:425-429, 1979.

372. Vizi ES, Volbekas V. Inhibition by dopamine of oxytocin release from isolated posterior lobe of the hypophysis of the rat: Disinhibitory effect of B-endorphin/enkephalin. Neuroendocrinology 31:46-52, 1980.

373. Vorherr H, Vorherr UF. Effect of prostaglandin (F_2, E_1 and E_2) on blood pressure and oxytocin-induced intramammary pressure responses in rats. Endocrinology 104:989-995, 1979.

374. Vorherr H, Vorherr UF, Solomon S. Contamination of prolactin preparations by antidiuretic hormone and oxytocin. Am J Physiol 234:F318-F324, 1978.

375. Wakerley JA, Poulain DA, Brown D. Comparison of firing patterns in oxytocin- and vasopressin-releasing neurone during progressie dehydration. Brain Res 148:425-440, 1978.

376. Wakerley JB, O'Neill DS, Ter Haar MB. Relationship between the

suckling-induced release of oxytocin and prolactin in the urethane-anaesthetized lactating rat. J Endocrinol 76:493-500,1978.

377. Walter R. Indentification of sites in oxytocin involved in uterine receptor recognition and activation. Fed Proc 36:1872-1878, 1977.

378. Walter R, Skala G, Smith GW. [5-aspartic acid]-oxytocin: First 5-position neurohypophyseal hormone analogue possessing significant biological activity. J Am Chem Soc 100:972-973, 1978.

379. Walter R, Smith CW, Roy J. Importance of the third amino acid residue of oxytocin for its action on isolated rat uterus: study of relationship between hormone conformation and activity. Proc Natl Acad Sci 73:3054-3058, 1976.

380. Walter R, Stahl GL, Caplaneris T, Cordopatis P, Theodoropoulos D. Active-site studies of neurohypophyseal hormones: synthesis and pharmacological properties of [5-(N^4, N^4-Dimethylasparagine)] oxytocin. J Med Chem 22:890-893, 1979.

381. Walter R, Wyssbrod HR, Glickson JD. Conformational studies on [3-D-alanine]-oxytocin and [4-D-alanine]-oxytocin in dimethyl sulfoxide by ^{1}H nuclear magnetic *resonance* spectroscopy. Interpretation in terms of a β-turn in the cyclic moeity. J Am Chem Soc 99:7326-7332. 1977.

382. Wanner O, Cranshaw DJ, Pliska V. The use of dynamic models to study the role of calcium in the oxytocin-induced contractions of the uterus. Mol Cell Endocrinol 6:281-292, 1977.

383. Ware S. Transplacental hyponatremia due to oxytocin. Br Med J 1:362, 1978.

384. Watkins WB, Choy VJ. Immunocytochemical study of the hypothalamo-neurohypophyseal system. III localization of oxytocin- and vasopressin-containing neurones in the pig hypothalamus. Cell Tiss Res 180:491-503,1977.

385. Watkins WB, Choy VJ. Maturation of the hypothalamo-neurohypophysial system II Neurophysin, vasopressin and oxytocin in the median eminence of the developing rat brain. Cell Tissue Res 197:337-346, 1979.

386. Watkins WB, Small CW, Walter R. Inactivation of neurohypophyseal hormones by colostrum and serum of human and other mammals. Pharmacol Res Commun 8:91-102, 1976.

387. Webb MT, Petrie RH, Pippenger CE. Fetal circulatory collapse during induction of labour in pregnant patient with epilepsy. Am J Obstet Gynecol 130:727-728, 1978.

388. Weindl A, Sofroniew MV. Neurohormones and circumventricular organs. An immunohistochemical investigation pp 117-137, in *Brain-Endocrine Interaction. III. Neural Hormones and Reproduction.* eds. DE Scott, GP Kozlowski, A Weindl, Karger, Basel, 1978.

389. Weingold AB, Yonekura ML, O'Kieffe J. Stress and nonstress antepartum fetal monitoring current status. Obstet Gynecol Annu 9:139-167, 1980.

390. Weinstein L. Oxytocin Challenge test. Am J Obstet Gynecol 132: 703, 1978.

391. Weitzerbin J, Goudea H, Gary-Bobo CM. Influence of membrane polarization and hormonal stimulation on the action of lanthanum on frog skin sodium permeability. Pfluegers Arch 370:145-153, 1977.

392. Weitzman RE, Glatz TH. Plasma oxytocin values. Am J Obstet Gynecol 132:913, 1978.

393. Weitzman RE, Glatz TH, Fisher DA. The effect of hemorrhage and hypertonic saline upon plasma oxytocin and arginine vasopressin in conscious dogs. Endocrinology 103:2154-2160, 1978.

394. Weitzman RE, Firemark HM, Glatz TH, Fisher DA. Thyrotropin-releasing hormone stimulates release of arginine vasopressin and oxytocin *in vivo*. Endocrinology 104:904-907, 1979.

395. Whalley ET. The action of bradykinin and oxytocin on the isolated whole uterus and myometrium of the rat in oestrous. Br J Pharmacol 64:21-28, 1978.

396. Whaley ET, Raynes G. Are prostaglandins involved in the action of bradykinin and oxytocin on the rat uterus *in vitro* and *in vivo?* J Endocrinol 81:134P-135P, 1979.

397. Whitfield MF, Salfield SAW. Accidental administration of syntometrine in adult dosage to the newborn. Arch Dis Child 55:68-70, 1980.

398. Whitton PD, Rodrigues LM, Hems DA. Stimulation by vasopressin, angiotensin and oxytocin of gluconeogenesis in hepatocyte suspensions. Biochem J 176:893-898, 1978.

399. Wildemeersch DA, Schellen AMCM. Double-blind trial of prostaglandin $F_{2\alpha}$ and oxytocin in the induction of labour. Curr Med Res Opin 4: 263-266, 1976.

400. Williams KI, El Tahir KEH. Effects of uterine stimulant drugs on prostacyclin production by the pregnant rat myometrium I. Oxytocin, bradykinin and $PGF_{2\alpha}$. Prostaglandins. 19:31-38, 1980.

401. Wilson PD. A comparison of four methods of ripening the unfavourable cervix. Br J Obstet Gynaecol 85:941-944, 1978.

402. Wilson WB. Midtrimester abortion with urea, prostaglandin $F_{2\alpha}$, laminaria and oxytocin. Obstet Gynecol 51:699-701, 1978.

403. Woodhouse DR. Water intoxication associated with high dose oxytocin infusion. Med J Aust 1:34, 1980.

404. Woolfson J, Steer PJ, Bashford CC, Randall NJ. The measurement of uterine activity in induced labour. Br J Obstet Gynaecol 83:934-937, 1976.

405. Wyssbrod HR, Ballardin A, Schwartz IL, Walter R,Van Binst G, Gibbons WA, Agosta WC, Field FH, Cowburn D. Side chain torsional angles and rotational isomerism of oxytocin in aqueous solution. J Am Chem Soc 99:5273-5276, 1977.

406. Yamaguchi K, Akaishi T, Negoro H. Effect of estrogen treatment on plasma oxytocin and vasopressin ovariectomized rats. Endocrinol Jpn 26: 197-205, 1979.

407. Yao AC, Nergardh A, Boreus LO. Influence of oxytocin and meperidine on the isolated human umbilical artery. Biol Neonate 29:333-342, 1976.

408. Zimmerman EA. Localization of hormone secreting pathways in brain immunocytochemical and light microscopy: a review. Fed Proc 36:1964-1967, 1977.

409. Zimmerman EA, Antunes J, Carmel PW, Defendini R, Ferin M. Magnocellular neurosecretory pathways in the monkey. Immunohistochemical studies of the normal lesioned hypothalamus using antibodies to oxytocin, vasopressin, and neurophysins. Trans Am Neurol Assoc 101:16-19, 1976.

410. Zimmerman EA, Defendini R. Hypothalamic pathways containing oxytocin, vasopressin and associated neurophysins. pp 22-29 in *Neurohypophysis*, eds. Moses Am, Share L, Basel, Karger, 1977.